Technology, Theory, and Practice in Interdisciplinary STEM Programs

Reneta D. Lansiquot
Editor

Technology, Theory, and Practice in Interdisciplinary STEM Programs

Connecting STEM and Non-STEM Approaches

Editor
Reneta D. Lansiquot
City University of New York
New York City College of Technology
Brooklyn, New York, USA

ISBN 978-1-137-56738-3   ISBN 978-1-137-56739-0 (eBook)
DOI 10.1057/978-1-137-56739-0

Library of Congress Control Number: 2016941866

Cover illustration: Abstract Bricks and Shadows © Stephen Bonk/Fotolia.co.uk

Printed on acid-free paper

This Palgrave Macmillan imprint is published by Springer Nature
The registered company is Nature America Inc. New York

To all those who explore what interdisciplinary studies add. Please add more to the world.

CONTENTS

NOTES ON CONTRIBUTORS

Bonne August is Provost and Vice President for Academic Affairs at New York City College of Technology, City University of New York (CUNY). Previously, she was Chair of the English Department and professor of English at Kingsborough Community College, CUNY. Dr. August has served as Principal Investigator for two institutional projects funded by the National Science Foundation (NSF). ADVANCE IT-Catalyst (2008–2011) addressed issues facing women faculty in STEM. The NSF I grant, The City Tech Incubator: Interdisciplinary Partnerships for Laboratory Integration (2009–2014) sought to generate both external and internal partnerships to strengthen students' hands-on experiences and research. Dr. August's professional interests and publications include work on poetry, women writers, writing assessment, portfolio assessment, and faculty development.

Mark A. Boyer is a University of Connecticut (UConn) Board of Trustees' Distinguished Professor of Political Science and serves as the director of the Environmental Studies program at UConn. His area of expertise is international affairs and environmental policy. He has published six books on global politics and presented numerous papers at international and national conferences. Currently, he serves as the Executive Director of the International Studies Association, the leading professional organization in his field. In addition, he was one of the original creators of the first iteration of the GlobalEd project and is a senior scientist on the current GlobalEd 2 Project.

Pamela Brown is Associate Provost at New York City College of Technology, City University of New York. She earned a PhD in Chemical Engineering from Polytechnic University, an MS in Chemical Engineering Practice from MIT, and a BS in Chemistry from the University at Albany, SUNY. Dr. Brown has published in journals ranging from *AIChE* to *JCE*. The Principal Investigator or Co-PI on five NSF grants totaling over $3 million, she has also served as Program Director at the National Science Foundation's Division of Undergraduate Education. She currently serves on the National Research Council of the National Academy of Sciences Committee, "Barriers and Opportunities in Completing Two and Four Year STEM Degrees," which is preparing a consensus paper for early 2016 release.

Scott W. Brown is a University of Connecticut Board of Trustees' Distinguished Professor of Educational Psychology. He has authored three books and more than 150 journal articles and book chapters, as well as presenting at numerous regional and international conferences. His research focuses on memory, problem-based learning, athletics, medical education, and educational technology. He is a Fellow in the American Educational Research Association and the Association for Psychological Science, and has received grants from the U.S. Department of Education, National Science Foundation, the Centers for Disease Control, and the Carnegie Corporation. He currently serves as the co-principal investigator on the GlobalEd 2 Efficacy Trials grant from the U.S. Department of Education's Institute for Education Sciences.

Elaine Correa is an associate professor in the Division of Education at Medaille College, in Buffalo, New York. She has taught in the fields of Women's Studies, Education, and Canadian Studies in the United States and Canada. Her research interests are in service learning, critical feminist pedagogy, popular culture, multi-culturalism, law, and e-learning. She is a strong advocate for experiential learning, designing opportunities for students to merge theory with practice. She has worked on several international research projects. She is currently the President of the Faculty, as well as serving as a mentor for her students at both the graduate and undergraduate levels.

Costanza Eggers-Piérola has thirty years of experience as an educator, researcher, developer, and educational consultant focusing on using data to transform teaching and learning. Her work as a Principal Investigator on various national and international projects is centered on researching and promoting equity as well as quality educational opportunities for all. Initiatives spearheaded by Dr. Eggers include several programs to stimulate statewide and institutional reform in science, technology, engineering, and mathematics (STEM) education. She has developed projects and programs using hands-on, inquiry-based curricula and has been awarded various federal, state, and private grants to support her Hispanic/Latino-serving initiatives. Dr. Eggers also contributes regularly as advisor, consultant, and professor of research, equity, and human rights in different higher education institutions in Spain.

Kevin P. Furlong is Professor of Geosciences at the Pennsylvania State University. He has over 110 refereed publications and is a Fellow of the American Association for the Advancement of Science and the Geological Society of America. His expertise is in plate tectonics, earthquake seismology, thermal geophysics, and natural hazards. He has been recognized at Penn State for his research and teaching contributions, and his accolades include the Eisenhower Award for Excellence in Teaching, the highest university award for teaching presented at Penn State. In addition, he has received 26 grants from the National Science Foundation and seven grants from other US government agencies. Moreover, he has received grants from, and consulted with, the energy industry, as well as serving on numerous international review panels.

Phil Ice is Vice President of Research and Development, American Public University System. He has conducted over 100 presentations and workshops, and he has authored more than 20 articles, book chapters, and white papers. His research focuses on emerging technologies in online learning environments. His work has been recognized internationally in the form of three Sloan-C Effective Practice Awards, the Gomory Award, the Innovation on Online Distance Learning Award, and the Adobe Higher Education Leaders Impact Award. He is a member of Adobe's Education Leaders Group and Higher Education Advisory Board. His vision for the future of technology is demonstrated by his inclusion on the NMC Horizon Report advisory board, and his role as Principal Investigator on a million-dollar WICHE/WCET grant exploring online retention patterns.

Reneta D. Lansiquot is Associate Professor and Program Director of the Bachelor of Science in Professional and Technical Writing at New York City College of Technology of the City University of New York, where she earned an AAS in Computer Information Systems and a B.Tech in Computer Systems. She earned an MS in Integrated Digital Media at Polytechnic University and a Ph.D. in Educational Communication and Technology at New York University. Her research focuses on interdisciplinary studies. Her first book is entitled *Cases on Interdisciplinary Research Trends in Science, Technology, Engineering, and Mathematics: Studies on Urban Classrooms.* This new book, *Technology, Theory, and Practice in Interdisciplinary STEM Programs: Connecting STEM and Non-STEM Approaches*, is being released simultaneously with *Interdisciplinary Pedagogy for STEM: A Collaborative Case Study.*

Kimberly A. Lawless is a Professor of Educational Psychology and the Associate Dean for Research at the University of Illinois at Chicago College of Education. She has published over 125 articles and book chapters, and presented her research at international, national, and regional conferences. Her research focuses on learning in technology-rich environments in K-12, post-secondary, and professional environments in a variety of disciplines. Her research has been supported by fed eral grants in excess of 30 million dollars, focusing on teacher professional development, reading and literacy, educational technology, the GlobalEd 2 project, and medical education. She currently serves as co-principal investigator on the GlobalEd 2 Efficacy Trials grant from the US Department of Education's Institute for Education Sciences.

Melissa Layne is Director of Research Methodology, American Public University System. She has written well over 30 peer-reviewed journal publications and 5 book chapters, and she has given 35 presentations and invited keynote speeches. Her research agenda includes topics on student retention, adaptive and personalized learning, multi-user virtual environments, self-paced instructional design and implementation, text analytics, informal learning, and quality assurance in online learning at the institutional, program, and course levels. Her research has been recognized by several distance learning organizations, including the National University Technology Network and the Distance Learning Administration organization. Layne also serves on the advisory council for the New Media Consortium, which is responsible for the annual issue of *The Horizon Report*. She also serves as Editor-in-Chief for *Internet Learning*.

Cinda P. Scott joined New York City College of Technology as the NSF I³ Program Manager and Coordinator of Integrated STEM Projects in November of 2010. As the City Tech I³ Incubator project coordinator, she oversaw the college's effort to provide City Tech students with a more integrated, cross-disciplinary laboratory experience. The focus of her work was on providing various departments with data to begin evidence-based transformation of STEM laboratories. Dr. Scott currently serves as Director for The School for Field Studies, Tropical Island Biodiversity Center in Bocas del Toro, Panamá, where she continues to synthesize her scientific expertise with her interest in the science of STEM education by working closely with faculty and students in the field.

Priya Sharma is Associate Professor of Education at the Pennsylvania State University. She has published numerous articles and book chapters, as well as presenting at national and international conferences. Her research focuses on the design and evaluation of technology-enhanced learning spaces, including face-to-face, online, and blended spaces. Her current work centers on learning with ubiquitous computing devices and learning within formal and informal online social networks. She has been the recipient of three grants from the National Science foundation related to technology, learning, and design.

LIST OF TABLES

CHAPTER 1

Introduction: Designing and Implementing Interdisciplinary Programs

Reneta D. Lansiquot

Abstract Interdisciplinary pedagogy and learning foster the cross-fertilization of ideas from different fields and disciplines, team collaboration across disciplines, the exploration of topics that reside at the boundaries and the edges of multiple disciplines, and the bringing of people together from various fields to explore issues and problems that have wide-ranging social impact. Chronicling the creation of an interdisciplinary Bachelor of Science in Professional and Technical Writing program at New York City College of Technology of The City University of New York, the chapter discusses practical matters of administration, such as choosing and integrating disciplines for a specialization. More conceptually, it examines the development of interdisciplinary programs through a focus on how technology, theory, and practice that connects STEM and non-STEM approaches in these programs.

Keywords Geosciences • Interdisciplinary studies • Interdisciplinary programs • Professional writing • Technical writing • Usability

R.D. Lansiquot (✉)
English, New York City College of Technology, City University of New York, Brooklyn, NY, USA

© The Author(s) 2016
R.D. Lansiquot (ed.), *Technology, Theory, and Practice in Interdisciplinary STEM Programs*,
DOI 10.1057/978-1-137-56739-0_1

1

This book investigates interdisciplinary programs in higher education in general, and connecting technology specifically, so as to promote interdisciplinary understanding[1] in science, technology, engineering, and mathematics (STEM) fields through a focus on how these developing programs function by examining the ways in which interdisciplinary teaching and learning can work in multiple fields. STEM-related interdisciplinary understanding is particularly vital among students majoring in science, technology, engineering, and mathematics, who often perceive that courses in their majors are not related to the general education (i.e., liberal arts and sciences) courses required for their degree. This separation prevents the transfer of skills between their general education courses and their degree pursuits.

The false dichotomy is particularly unfortunate, because solving the daunting challenges of the twenty-first century—such as drug-resistant bacteria, the scarcity of natural resources, and climate change—requires global citizens armed with robust, complex abilities who can integrate interdisciplinary concepts with bold technologies. Perhaps the most promising way to promote the sort of thinking in which our learners transfer knowledge between courses, across disciplines, and among research fields is through interdisciplinary studies, which have been found to facilitate problem solving in numerous studies.[2] Thus, when developing the professional and technical writing degree program at New York City College of Technology (City Tech) of the City University of New York, I wanted to make sure that students were engaged in interdisciplinary writing as problem solving.[3] As part of this program, to provide depth in a content area, students must complete a series of courses in a single professional, scientific, or technical discipline, which account for at least 15% of their required courses. Students who transfer from other programs (e.g., from an Associate in Arts in Communication Studies or an Associate in Applied Science in Health Information Technology) and those with different backgrounds, such as in educational technology, support the interdisciplinary nature of the program and provide a robust exchange of ideas.

My seven-year journey while creating this interdisciplinary professional and technical writing undergraduate degree program included overcoming the challenges of framing this program and negotiating pedagogical and political terrain of the various specializations. Now, this new program and its specializations (currently, Architectural Technology, Biology, Chemistry, Communication Design, Computer Science, Public Health, Economics, Psychology, and Social Science) reveal how an interdisciplinary program can work. While I was creating the program, deciding on the courses to be included in each specialization meant meeting with department Chairs and Discipline Coordinators—for instance, because Economics, Psychology, and Social Science are all housed in the Social Science Department, I had to meet with that Chair and Discipline Coordinators. Before the meetings, I found it invaluable to explain the intent of the specialization and providing a tentative list of courses that took into account prerequisites and the level of difficulty for students who were not majoring in the specialization; the meetings were then framed around a discussion of suggested changes. My task was made somewhat easier because our Public Health specialization is itself interdisciplinary, including courses in government, public policy, health, and human services.

City Tech's Bachelor of Science in Professional and Technical Writing program prepares students to communicate clearly and effectively using a variety of tools and media. Students learn how to translate complex, industry-specific information into lay terminology or another industry-specific discourse. In order to meet the needs of industry, the program allows students to look across disciplinary boundaries, bringing together information and skills from a variety of fields into a new base for learning, designing, and writing. The structure of this degree ensures that students who graduate from this program (a) master industry standard applications for professional and technical writing and related technologies, (b) acquire expertise in a professional studies-related, science-related, or technology-related discipline that will give them an edge in the marketplace, and (c) enter a rapidly shifting workplace prepared to negotiate new forms of media with sophistication and confidence. The program provides students with both a hands-on experience using a range of tools as well as an understanding of the theories underlying the use of those tools. Graduates master industry standards for both professional and technical writing, as well as related technologies.[4]

Some of the overarching concepts informing the interdisciplinary approach to learning are discussed in Chap. 2. Melissa Layne and Phil Ice examine the elements of the emerging platform and analytic technologies, emphasizing their potential impact on the three Community of Inquiry (CoI) presences: teaching, social, and cognitive. Attention is given to technologies that may have positive or negative impacts on collaborative, constructivist interdisciplinary learning models.

In Chap. 3, Priya Sharma and Kevin P. Furlong report on a multi-year collaborative research effort between the geosciences and education, focusing on the design and evaluation of modules that engage undergraduate students in science reasoning skills. They discuss their design-based research approach to constructing active learning modules to engage students in a large general education undergraduate course in natural hazards, as well as their approach to integrating mobile devices in an upper-level undergraduate course on the same topic. Overall, the chapter identifies the features of interdisciplinary courses that support scientific reasoning, student collaboration, and technology-enhanced learning in an undergraduate classroom.

In Chap. 4, Kimberly A. Lawless, Scott W. Brown, and Mark A. Boyer continue the exploration of collaboration, problem solving, and reasoning in an interdisciplinary context. They point out that across several independent surveys of businesses and potential employers, the most commonly cited skills that industry requires in newly graduated college students include the abilities to solve complex, multidisciplinary problems, work successfully in teams, exhibit effective oral and written communication skills, and practice good interpersonal skills. However, industry leaders point out that many students who obtain their postsecondary degrees do not possess these skills, and as such are not fully prepared to successfully participate in the twenty-first century workforce.[5] To address this need, they designed the GlobalEd 2 (GE2) program, which engages classrooms of students in online, simulated negotiations of international agreements on issues of global concern, such as water scarcity and climate change. Their GE2 program is an interdisciplinary problem-based curriculum targeting students' global awareness, scientific literacies, and twenty-first century workforce skills. Their results over the past 15 years using various iterations of GE2 have been implemented in classrooms, ranging from middle school through college, demonstrate the positive impact of GE2 along a number of dimensions including writing, argumentation, science knowledge, and social-perspective taking. This chapter provides an overview of GE2, its design principles, and discusses the data from a recent implementation with college freshmen, specifically focusing on gains with respect to self-efficacy across multiple domains.

While Lawless, Brown, and Boyer explain how interdisciplinary courses can make students better citizens of the world, Elaine Correa explains how these courses can make students better citizens in their communities. She discusses how the integration of interdisciplinary studies with service learning can invoke meaningful, engaging, and sustainable learning with technology for students beyond the classroom. Service learning provides learners with opportunities to explore the real needs of a community, connecting content knowledge with prior experiences from both the classroom and life. Grounded in Dewey's notion of "learning by doing,"[6] service learning necessitates deep reflection as students merge theory with practice. In Chap. 5, she explains how an interdisciplinary studies approach to service learning offers space wherein the adage still applies for students today, "Tell me, and I will forget. Show me, and I may remember. Involve me, and I will understand."

It is with this adage in mind that *The City Tech I³ (Innovation through Institutional Integration) Incubator: Interdisciplinary Partnerships for Laboratory Integration*, a program supported by the National Science Foundation (NSF), was created, integrating research and education with a focus on inquiry as a means of learning and the development a global workforce by expanding industry partnerships to provide real life application of STEM learning. This project is a catalyst for transforming laboratory curricula and teaching across STEM departments by establishing innovations that will invigorate courses and lead to a greater integration of STEM projects across the college. Several multiyear projects totaling more than three million dollars benefitted from this program, generating ongoing and broad-reaching changes across STEM programs and student services. This chapter explores the program features that contributed to the faculty's understanding and teaching of STEM and the nature of policies, procedures, and partnerships that supported the effectiveness of the program. It highlights transformative approaches to recruitment, teaching, mentoring, supervision, and communication and collaboration within and across laboratories. Innovations in these areas were intentionally spread from one lab to another and intentionally institutionalized within the college and laboratory culture. This chapter also contributes to the dialogue on best institutional approaches focused on attracting, retaining, and preparing a diverse student population in STEM fields. The cross-institutional strategies, faculty development, and initiatives described in the article provide real life examples of what works towards these goals and what sustains and multiplies these efforts. More than a dozen NSF projects at City Tech were leveraged to enhance cross-institutional communication and collaboration, provide synchronicity in goals and strategies, and ensure continuity of the goals of the program.

These projects have helped transform undergraduate education, and the contributing authors of this book take multiple perspectives on the interdisciplinary studies in the different institutions, with a particular focus on connecting technology to educational theory and learning practice. These chapters explore interdisciplinary online collaborative learning, design methodologies, the designing of technology-enhanced active learning environments, STEM literacy, an interdisciplinary approach to service learning, and strategic institutional integration. The goal of this book, *Technology, Theory, and Practice in Interdisciplinary STEM Programs: Connecting STEM and Non-STEM Approaches*, is to provide innovative interdisciplinary studies for a scholarly community. The diversity in pedagogy presented here reflects the intended beneficiaries of this book: our demographically diverse students, who are preparing to enter a world where their problem-solving skills are much needed.

Notes

1. Veronica Boix-Mansilla and Howard Gardner, "Teaching for Understanding—Within and Across the Disciplines," *Educational Leadership* 51, no. 5 (1994): 14–8.
2. Susan Elrod and Mary J. S. Roth, "Framing Leadership for Sustainable Interdisciplinary Programs," *Peer Review* 17, no. 2 (2015): 8–12; Lisa R. Lattuca, *Creating Interdisciplinarity: Interdisciplinary Research and Teaching among College and University Faculty* (Nashville, TN: Vanderbilt University Press, 2001); Lisa R. Lattuca, Lois J. Voigt, and Kimberly Q. Fath, "Does Interdisciplinarity Promote Learning? Theoretical Support and Researchable Questions," *Review of Higher Education* 28, no. 1 (2004): 23–48; Project Kaleidoscope, *What Works in Facilitating Interdisciplinary Learning in Science and Mathematics: Summary Report* (Washington, D.C.: AAC&U, 2011).
3. See Linda Flower, *Problem-Solving Strategies for Writing*, 4th ed. (New York: Harcourt Brace Jovanovich, 1993).
4. For more information see http://www.citytech.cuny.edu/academics/deptsites/english/degrees.aspx.
5. American Association of Colleges and Universities (AAC&U), "Raising the Bar: Employers' Views on Colleges Learning in the Wake of the Economics Downturn," 2010, https://www.aacu.org/sites/default/files/files/LEAP/2009_EmployerSurvey.pdf (accessed May 11, 2015); Gallup Postsecondary Education Aspirations and Barriers, 2014, http://www.luminafoundation.org/resources/postsecondary-education-aspirations-and-barriers (accessed on August 31, 2015); John Morely, "Labour market developments in the new EU Member States," *Industrial Relations Journal* 38 no. 6 (2007): 458–79. Also refer to Chap. 4 of the present book.
6. John Dewey, *Experience in Education* (New York: Macmillan, 1939).

BIBLIOGRAPHY

American Association of Colleges and Universities (AAC&U). 2010. Raising the bar: Employers' views on colleges learning in the wake of the economics downturn. Accessed August 31, 2015. https://www.aacu.org/sites/default/files/files/LEAP/2009_EmployerSurvey.pdf.

Boix-Mansilla, Veronica, and Howard Gardner. 1994. Teaching for understanding—Within and across the disciplines. *Educational Leadership* 51(5): 14–18.

Dewey, John. 1939. *Experience in education.* New York: Macmillan.

Elrod, Susan, and Mary J.S. Roth. 2015. Framing leadership for sustainable interdisciplinary programs. *Peer Review* 17(2): 8–12.

Flower, Linda. 1993. *Problem-solving strategies for writing,* 4th ed. New York: Harcourt Brace Jovanovich.

Gallup. 2014. Postsecondary education aspirations and barriers. Accessed June 9, 2015. http://www.luminafoundation.org/resources/postsecondary-education-aspirations-and-barriers.

Lattuca, Lisa R. 2001. *Creating interdisciplinarity: Interdisciplinary research and teaching among college and university faculty.* Nashville: Vanderbilt University Press.

Lattuca, Lisa R., Lois J. Voigt, and Kimberly Q. Fath. 2004. Does interdisciplinarity promote learning? Theoretical support and researchable questions. *Review of Higher Education* 28(1): 23–48.

Morely, John. 2007. Labor market developments in the new EU member states. *Industrial Relations Journal* 38(6): 458–479.

Project Kaleidoscope. 2011. *What works in facilitating interdisciplinary learning in science and mathematics: Summary report.* Washington, DC: AAC&U.

CHAPTER 2

Emerging Technologies and Potential Paradigmatic Shifts in the Community of Inquiry Framework

Melissa Layne and Phil Ice

Abstract Despite extensive Community of Inquiry (CoI) research, dependence on the technological foundations of online learning via the Learning Management System (LMS) remains underresearched. As a blank canvas for course creation, the LMS is a collection of tools and communication devices through which content and activities are developed. Instructional designers follow a linear design pathway where there is little potential for deviation; however, the ease of course development typically outweighs some LMS limitations. Although online learning and the CoI have evolved, advancements in learning technologies offer LMS alternatives. This chapter examines emerging platform and analytic technologies, emphasizing their potential impact on the three CoI presences. Attention is given to technologies that may have positive or negative impacts on collaborative, constructivist interdisciplinary learning models.

Keywords Community of inquiry • Emerging technologies • Learning management system • Learning theories • Paradigm

M. Layne (✉) • P. Ice
American Public University System, Charles Town, WV, USA

© The Author(s) 2016
R.D. Lansiquot (ed.), *Technology, Theory, and Practice in Interdisciplinary STEM Programs*,
DOI 10.1057/978-1-137-56739-0_2

9

Moving from a curiosity that was utilized by a few science departments to a tool that could provide a viable means for instantaneous communication, the Internet began to garner the attention of academics in the early to mid 1980s. Although reliant upon fundamental tools, such as bulletin board services, it became clear that a new tool was emerging, in which ideas could be interacted upon in real time, regardless of location. Soon thereafter the first pioneering efforts in online learning began to emerge. While still largely a curiosity, these early initiatives demonstrated that learning could be unbundled from the universities' traditional brick and mortar confines.

In 1993, the National Center for Supercomputing Applications released the first Internet browser, Mosaic, which was followed a year later by Netscape's entry into the market as a commercial entity. Although primitive by today's standards, these early appliances allowed a richer browsing experience that seamlessly incorporated text and images to be simultaneously displayed, as well as rudimentary scripts to run applications such as forums, e-commerce, and so forth. Shortly thereafter, pioneering efforts to create richer learning experiences followed. However, the development of significant bodies of learning materials was still quite complicated and required a deep understanding of coding and web design to create all but the simplest of artifacts.

In 1995, Professor Murray Goldberg, of the University of British Columbia, began developing the first browser-based Learning Management System (LMS), which was released under the name WebCT in 1996 and commercialized the following year. Intended to simplify the process for putting learning materials and activities online, WebCT offered an interface that allowed users to upload documents rapidly, create simple web pages, conduct quizzing activities, participate in threaded discussions, and interface with institutional enterprise systems. For those engaged in early online course development, WebCT and its successor, LMSs, offered a quantum leap forward for the overwhelming majority of academics and catalyzed online learning. However, the tradeoff for the ability to rapidly create online courses was that the resultant product would be constrained.

As with most technological innovations that emphasize simplicity, users understand that creativity is constrained. Although conformity is usually considered antithetical to the tenants of higher education, this is not the case for users of new technology. When interacting with technology, the vast majority of faculty, who are experts in their own fields, are beginners and seldom wish to acquire advanced skills. The role of faculty in institutions of higher education has traditionally been one that is defined by achievement in teaching, research, and service, and thus faculty members are often reluctant to spend the time and effort to learn new skills unless they can be incorporated into the research/teaching/service trinity. Further, in the late 1990s, virtually none of the institutions were amenable to such role modifications, although there is more acceptance of the need for acquisition of such skills today.

The skills needed to succeed in the online educational environment are not limited to learning software. Traditionally, teaching has been thought to demand cognitive, affective, and social competencies. The cognitive aspect consists of those acts that foster the conveyance of knowledge. Affective aspects are related to those roles assumed by the instructor that influence the relationships within the classroom setting as well as the external but related relationships, such as mentoring and support, that are formed between students and instructor. Finally, the social functions related to the teaching component consist of policy enforcement and conflict resolution.

However, when instructors are asked to develop and deliver online courses, conflicts often arise with the traditional rubrics by which performance is assessed. Berliner notes that, when using new technology, many teachers revert to novice status.[1] For the instructor who, through years of practice, has developed a teaching style that allows him or her to teach in a seamless, fluid manner, this reversion can often lead to the belief that technology-mediated learning is inferior to the traditional mode in which they are well versed. Specifically, the act of teaching can no longer be easily defined in terms of cognitive, affective, and social roles. When moving to an online teaching environment, clearly defined managerial and gate-keeping roles emerge from the traditional cognitive and affective aspects of teaching. As such, faculty embracing the new technology acquiesced to a *quid pro quo*, sacrificing originality in the name of expedience.

Following the highly templatized models afforded by the LMS, mass assembly of online courses began in earnest within a few years of the introduction of the first platforms, and they have continued to expand at a rapid rate until present. During this time, many new LMSs have entered the marketplace and older platforms have expanded with claims of enhanced functionality continually extolled. However, compared to technologies present in the commercial space, educational technologies, especially LMS providers, have only demonstrated incremental advancement. Indeed, it is fair to say that from a technical perspective, the framework of the LMS has remained nearly stagnant, with only cosmetic front-end enhancements.

In their 2014 EDUCAUSE report, *The Current Ecosystem of Learning Management Systems in Higher Education: Student, Faculty, and IT Perspectives*, Dahlstrom, Brooks, and Bichsel[2] present an interesting look at the higher education market penetration of the leading learning management systems from 2005 through 2009, with Blackboard continuing to take the lead. Twenty years after the humble beginnings of the first learning management systems, one can clearly see the presentation of the same LMS providers from 2005 to 2009. The EDUCAUSE report corroborates and substantiates the claim that LMS use by US higher education institutions has remained constant.[3]

Outside of higher education, the LMS has played an important part in corporate training, teaching, and learning in K-12 schools. Despite the appearance of stability in the above mentioned reports, it should also be noted that statistics by the K-12 and corporate sectors are embedded within these consistent numbers on LMS use. According to the EDUCAUSE Core Data Service (CDS) LMS revenue was reported at $1.9–2.6 billion in 2013, and is expected to increase to $7.8 billion by the year 2018.[4] Moreover, and most importantly, LMS has provided countless opportunities for students of all ages to attain an education online; without this important innovation, many would not have been able to attain an education otherwise. Over the years, the LMS—whether used by students, faculty, administrators, or corporate employees—has been considered ubiquitous, and has proven beneficial across all sectors.

As previously noted, this rapid market expansion has not necessarily translated into advancements in functionality. However, the tradeoff for the vast majority of practitioners more than offset any disadvantages. As such, the origins of mainstream online learning should be contextualized through the lens of the LMS. Specifically, with the widespread adoption of LMSs and the similarities in functionality between platforms, online pedagogy and learning theory have been guided by the confines of the LMS in a manner that parallels similar developments in traditional environments vis-à-vis physical arrangements and affordances.

Theoretical Underpinnings to Support Teaching and Learning Within the LMS

One of the most important things to consider when implementing any teaching and learning platform is, naturally, the customers. In academe, this translates to the students and those who will be developing activities, assignments, assessments, and so on—the course instructors. How students and faculty interact while working within these digital platforms is paramount; therefore, implementing best practices is an absolute must.

Shortly after the introduction of the WebCT LMS back in 1997, what quickly and simultaneously became apparent with this exciting, new environment was the necessity of a supporting teaching and learning theoretical framework. Because the introduction of the LMS swiftly turned the world of face-to-face education completely upside down, some early researchers who were anxious to research this new learning environment made the methodological mistake of comparing the two teaching and learning environments. Many scholars maintained that the already established, tried-and-true theoretical frameworks developed for face-to-face teaching and learning *should* and *would* also support teaching and learning within this new online environment. Unfortunately, as quickly as the need for a framework was acknowledged, not much later did researchers realize that existing theoretical frameworks would not serve as a good "fit" in this new medium. As much as it seemed logical to compare "apples to apples," scholars were challenged to (a) begin exploring student and faculty experiences within the LMS; (b) examine thoroughly and thoughtfully student and faculty interactions with the LMS tools and functions; and, (c) begin constructing pedagogy around explaining and supporting teaching and learning based upon these experiences. In essence, researchers found themselves returning to the drawing board to conceptualize a theory not yet conceived.

Learning Theories and Their Impact on Learning Environments (1950–2015)

Central to educational research and practice resides a basis from which, methods, strategies, instructional design, interactions, student engagement, and so forth can be explained and used to inform and validate teaching and learning practice. Specifically, theoretical frameworks serve as a guide on how to approach and effectively plant the seeds of knowledge among diverse student groups. Thus, it is imperative to understand and apply, not just a single theory, but a variety of theories. Oftentimes, instructors do not deviate from a preferred educational framework, one that presents the instructor as a fountain of knowledge. Unfortunately, this type of commitment is counterproductive; if it is consistently maintained throughout a child's educational trajectory, it eventually imparts to some the perspective that learning ends after high school (or earlier). Further, given the diversity of learners, effectively meeting students' needs and fostering an inherent desire for lifelong learning demands that instructors weave together situational aspects, multidimensional contexts and goals, and multiple educational theories.

Innovation and science, for example, will always continue to directly impact our societal, cultural, and educational landscape. Many of these changes occur gradually, while others are almost instant. Nevertheless, these changes cause us to reimagine, rethink, and reassess. For those of us who are involved in some aspect of education, we can almost count on quick changes without much warning. Therefore, the benefits are greater when being proactive rather than reactive. Sometimes this is difficult to do, but the more metacognitive we are with regard to the importance of understanding and applying appropriate theories into practice before these changes take place, the better off we will be—and even more importantly, our students will be better off.

Building a Learning Environment from Theory

Constructing learning environments is important to consider when assessing student learning. When closely examining various theories through an instructor's lens, there may initially exist some gray area in terms of how to create effectively an optimal learning environment based upon individual theoretical frameworks—and an even more gray area when there are multiple perspectives to take into account and/or a new technology tied to the learning environment. Let's take a look at (a) the theorists responsible for developing the most commonly implemented theories; (b) the theory each one developed; (c) the type of learning environment that emerged from the theory; and (d) the instructional technology that was used to represent and support the learning environment.

Behaviorism and the Overhead Projector

B. F. Skinner formulated the operant conditioning learning theory, behaviorism. Behaviorism is a worldview that operates on a principle of "stimulus-response." All behavior is caused by external stimuli (operant conditioning). All behavior can be explained without the need to consider internal mental states of consciousness. An example of this concept is illustrated in Skinner's "Teaching Machine" (a video of this is available at YouTube, https://youtu.be/EXR9Ft8rzhk). This machine provided assessment-centered standardized test questions to students. Upon providing the correct answer, the student received a piece of candy.[5]

Robert Gagné and Scantrons

The cognitivist paradigm essentially argues that the "black box" of the mind should be opened and understood. The learner is viewed as an information processor (like a computer).[6] People needed to be taught complex skills, and simply knowing what to do and doing it (basic S-R) does not determine success in all instances. Robert Gagné developed a theory of learning that accounted for the variety of human understandings with the conditions of learning. Gagne's theory posited five categories of learning: verbal information, intellectual skills, cognitive strategies, attitudes, and motor skills.[7] The skills to be learned are written into performance objectives, and the category of learning is identified. Cognitivist theory ultimately ushered in educational television programs.

Constructivism and Collaborative Activities

Constructivism is a paradigm or worldview that posits that learning is an active, constructive process. The learner is an information constructor. People actively construct or create their own subjective representations of objective reality. New information is linked to prior knowledge, thus mental representations are subjective. Constructivism promotes that the design of learning environments support the construction of knowledge by the learners.

Humanism and Discussion Boards

Humanism is a paradigm, a philosophical and pedagogical approach that believes learning is a personal act to fulfill one's potential.[8] Competition in the workforce between traditional and nontraditional students during this time prompted the revisiting of Knowles' Adult Learning theory as many adults headed back to the classroom to gain additional knowledge and skills. Adult Learning theory is a set of five assumptions on how adults learn and absorb knowledge: (1) Self-concept—the notion that a person's self-concept moves from dependence to self-directed; (2) Adult learner experience—the assumption that people continually gain knowledge from life experiences; (3) Readiness to learn—a person's readiness to learn adjusts to his or her developmental tasks and social roles; (4) Orientation to learning—a person's perspective of learning shifts from subject-centeredness to problem-solving; and (5) Motivation to learn—as people mature, they develop an internal desire/motivation to continually learn. Online learning has enabled more and more nontraditional students—many of whom hold full or part time jobs—to attain a postsecondary education. One component within the LMS, the discussion board, allows for asynchronous social communication between students and instructors, thus providing flexibility around demanding work schedules.

Twenty-First Century Skills and Virtual Learning Environments
Twenty-first century skills are skills deemed necessary for students to master in order for them to experience success in school and life in an increasingly digital and connected age; the competencies include digital literacy, traditional literacy, content knowledge, media literacy, and learning/innovation skills.[9] Virtual Learning Environments, such as *Second Life*, ushered in a more multidimensional option for teaching and learning online.

Adaptive Learners and Interactive Environments
These environments allow learners to engage themselves with learning content. Teaching and learning environments need to be more varied and flexible spaces where students are learning proactively.[10] We believe that there are some skills from the last decade that will continue to be important in addition to development of a new set of skills: (a) critical thinking; (b) collaboration/initiating and developing partnerships; (c) communication; (d) creativity; (e) adaptability and flexibility; and (f) self-discipline/regulation.

LMS Tools and the Paradigm Shift to Social Learning

Although Wilson and Peterson were referring to education in an overarching sense when they stated, "Perhaps the most critical shift in education in the past 20 years has been a move away from a conception of 'learner as sponge' toward an image of 'learner as active constructor of meaning',"[11] we believe that the introduction of online learning twenty years ago was largely responsible for inducing this significant shift. Up until this groundbreaking and game-changing innovation, "teaching" largely consisted of instructors positioned behind their podiums and spending the majority of class time presenting the topic *du jour* as meticulously outlined in the course syllabus. A student's role in the classroom was to simply listen to the lecture, take notes, and passively absorb information—much like the biological process of osmosis. In short, teachers talked, and students were expected to listen.[12] Needless to say, up until the emergence of the Internet, there was not much time allotted for group activities, collaborative projects, or collective brainstorming. Social and community-centered interactions were affordances that were not included within course instruction and therefore completely independent of most face-to-face classrooms.

An early dissenter from this paradigm was John Dewey, from the constructivist camp, who believed that individual and societal interests must meld together in order for an educational experience to occur.[13] His perspective of authentic learning is predicated upon the belief that when fused together, a learner's private and public experiences facilitate curiousness and an innate desire to inquire. Further, rooted in constructivist theory, Dewey asserted that inquiry is a social and collaborative activity.

In terms of online learning environments (such as the LMS) the "classroom" is a virtual space where students can learn both individually and collaboratively vis-à-vis tools such as discussion boards, chat tools, and so forth. Understanding the need for social presence in the classroom, researchers immediately pondered how this social aspect of learning and community-building would be established when learners were physically apart from each other. Early research conducted on this important component—within the confines of learning management system tools such as discussion boards—confirmed that social presence could indeed be developed in an online learning environment, but the much bigger question that lay ahead was the identification of specific social presence indicators. As instructors began experimenting with ways in which to facilitate social presence within discussion boards, in particular, they noticed that students were able to project their personalities within their self-constructed narratives.[14] Several research studies on the establishment of online social presence followed, examining both instructor and learner perceptions of these virtual connections between and among each other.[15]

Although the idea of two-way knowledge transfer through online social collaboration was not fully developed, by the late 1990s works such as Gunawardena's had made clear that a paradigm shift was clearly taking place.[16] Foremost among the factors taken into consideration is that fact that effective learning interactions occur when there are collaborative activities within the confines of course that are predicated upon cognitive scaffolding. The skills developed and the knowledge learned has practical application to future courses and life activities.

From a theoretical perspective, Green defines the acquisition of knowledge as a process in which facts, evidence, and beliefs interact and are modified by groups of learners and those providing the evidence and interpretations of the facts.[17] Freire defines this type of knowledge co-construction as problem posing because it requires continual input and modification at various levels to achieve consensus solutions to a given problem.[18] Using these two models as a foundation, it becomes possible to understand interaction and feedback as resulting both from the actions of those who initiate the learning activities, or create a problem-posing situation, and from the actions of the learners themselves.

A working model of this dualistic interpretation of interaction and feedback can be found in the principle of Chaordic Theory, which asserts that when given a catalyst for origination and only loosely established boundaries, individuals will continually solve problems and simultaneously create new ones.[19] Within this cycle, the roles of the individuals may change with respect to resolving a given problem, but their contribution to the overall goal is related to the group dynamic that emerges. In this type of a system, the process of goals being initially established by an instructor and interaction thereby initiated would be a function of the pedagogical and technical design elements of online course construction and delivery. The process in which groups of learners work with each other and the instructor to find solutions produces further degrees of interaction and continuous feedback loops, such as those envisioned by theorists and practitioners who were beginning to explore online learning during these formative years.

To explain these processes in the online context, Randy Garrison, along with researchers at the University of Calgary, looked to the Practical Inquiry Model, which was later renamed and incorporated in the Community of Inquiry Framework as Cognitive presence. The Practical Inquiry model is defined by two axes. The vertical axis reflects the integration of thought and action. This also emphasizes the collaborative nature of learning and the need for community. The integration of discourse and reflection (i.e., public and private worlds) is a key feature of this model. Although identified as two distinct processes, in practice these dimensions (i.e., discourse and reflection) are most often indistinguishable and instantaneous iterations. The horizontal axis represents the interface of the deliberation and action axis. The extremes of the horizontal axis are analysis and synthesis, which are the points of insight and understanding.[20]

The section that follows furthers explores and defines the Community of Inquiry and its component parts. This is followed by a discussion of tools that exist within the LMS, how these are evolving, and the impact on the model.

An Overview of the Community of Inquiry

Although several models have been proposed to explain the learning process in online environments, the one gaining the most attention is the Community of Inquiry Framework (CoI). Grounded in the constructivist school of thought, the CoI consists of three overlapping elements: (1) teaching, (2) social presence, and (3) cognitive presence. These three elements coalesce to create the educational experience. A search of Google Scholar revealing more than 500 citations, and numerous confirmatory analyses have been conducted by numerous researchers—thus, the CoI is considered a baseline for the establishment of grounded theory in online teaching and learning dynamics.

In the context of online learning, social presence is described as the ability to project one's self through media and to establish personal and meaningful relationships. The three main factors that allow for the effective projection and establishment of social presence are effective communication, open communication, and group cohesion. Grounded in the work of Dewey, cognitive presence is defined as the exploration, construction, resolution, and confirmation of understanding through collaboration and reflection. This process is described as consisting of four phases, beginning with creating a sense of puzzlement or posing a problem that piques learners' curiosity. As a community, course participants exchange information and integrate their understandings to answer the initial problem, culminating in the resolution phase, where learners are able to apply the knowledge to both course and non-course related issues. Teaching presence, the third component of the CoI, is described as a three-part structure: (1) facilitation of discourse, (2) direct instruction, and (3) instructional design and organization. The first element, facilitation of discourse, is necessary to maintain focus and engagement in course discussions. It also allows the instructor to set the appropriate climate for academic exchanges.

Despite the extensive amount of research concerning the CoI and its general acceptance in the field, one of the least-researched aspects of the construct is its grounding and dependence on the technological foundations of online learning, specifically the Learning Management System. When viewed as a blank canvas for course creation, the LMS can be conceptualized as a collection of fluid authoring tools and communication devices through which content and activities are developed. Depending upon the instructional design paradigm being utilized, courses are systematically constructed to move learners from a set of goals through a series of activities and culminating assessment activities. Notably, with the exception of a few emerging platforms, instructional designers must follow a linear pathway in which there is little potential for deviation. However, the ease with which courses can be designed and developed has almost always outweighed limitations inherent in LMS. In academia, this trade-off has been reinforced by the need to ensure student privacy by utilizing systems that can reside entirely behind firewalls, another attribute of the LMS. Online learning and the CoI have evolved against this backdrop. However, recent advancements in learning technologies offer alternatives to the LMS-centered learning experience. This chapter deconstructs emerging platforms and analytic technologies, emphasizing their potential impact on each of the three CoI presences. Special attention is given to those technologies that may have extremely positive or negative impacts on collaborative, constructivist interdisciplinary learning models.

Common LMS Tools and Functions

Although LMSs have evolved in various ways over time, what has made the LMS so intriguing to researchers, in particular, are common core tools specifically developed for various modes of online communication between and among multiple users. Not only are they are ideal environments for storing information and course content for student access, but they also offer effective ways to deliver, assess, and record grades from assignments, quizzes, and tests. For example, albeit not quite the novelty nowadays, the simple task of uploading, downloading, sending, and receiving digital correspondence/documents (either to remain digital, or to print) within a secure LMS environment is an important function stemming back to the beginnings of the Internet, and more specifically, email providers. Additional teaching, learning, administrative, assessment tools and their respective functions commonly found within current LMSs include, but are not limited to the following:

Announcements
Announcements are an online tool that is typically used by the instructor to post important information. Announcements are commonly found on the students' home pages upon logging into the LMS and can include multimedia, text, images, hyperlinks, and so on. Often used in conjunction with a calendar, announcements were the basis for the Instructional Design and Organization component of the CoI's Teaching Presence construct. It was here that expectations were set and students were briefed on upcoming events, changes to course assignments, and where the instructor provided brief thoughts on the week's or unit's assignments. However, as the LMS is housed behind a firewall, students have to go through a login process to access this information.

In other parts of their lives, students likely use Twitter, Facebook Messenger, text messages, and so forth to receive these types of updates. As such, accessing this type of service in a closed environment may actually have a negative impact on student performance as they may believe that it is an unnecessary burden to have to go through a login process to receive snippets of information that would otherwise be sent to them. In response, some LMS providers are starting to offer push services that allow for access to announcement information vis-à-vis the types of services previously mentioned.

Moving forward, it will be interesting to see what impact these push services will have on students' perception of the online classroom. Will they continue to view it as an isolated entity? Or will the definition of classroom be expanded? If the latter is true, we may see a strengthening of perceptions of teaching presence, but social presence indicators may decline as the concept of a classroom will become far more distributed.

Discussion Boards/Forums/Wikis
The discussion board is an online tool allowing for the asynchronous, communicative exchanges between/among multiple users within an LMS. For educational purposes, discussion boards have allowed for the sharing and receiving of information and/or opinions relating to a particular topic.

Originally designed to facilitate dialogue around a given topic within the course, the discussion forum provided a convenient means of achieving what was a rather complex piece of scripting in the early days of the LMS. However, there are now numerous stand-alone services that offer discussion board functionality which can be embedded in any webpage with a modicum of effort. In fact, it is hard to think of any service-oriented website that does not have this type of functionality for purposes of customer feedback and elaboration.

In its simplest manifestation, the discussion forum was one of the very earliest forms of social media, with remnants still visible in such services as Facebook. However, contemporary social media offers functionality that far exceeds that of the humble discussion forum. The downside is that many students do not want to have their personal online lives intertwined with academia, as has been discovered when numerous instructors have attempted to offer courses on Facebook, despite the considerable gains in Social Presence that might be achieved by doing so.

A solution that may offer a healthy compromise is Hoot.me (http://hoot.me). Hoot.me connects Facebook, where students are, with LMS providers like Canvas and Blackboard, where instructors are, to ensure that every question can be exposed to the widest possible audience. Hoot.me also has filters that allow users to define which networks, internal or external, that messages will be viewed on.

Beyond the ability to expand networking to find solutions to problems, services such as this dramatically expand the potential for the Exploration phase of Cognitive Presence, allow for deeper reflection in the Integration phase, and richer responses in Resolution. However, there are also a few potential downsides. One of the foundations of Teaching Presence is that the instructor should keep students' discussions on track and provide guidance where necessary. This may become much more difficult if networks expand and faculty do not participate. Secondly, there is the question of whether these expanded networks around course problems will impact coherence of the learning community and thus social presence.

Certainly, while this type of solution is exciting and offers many opportunities, there are likely to be numerous points at which the CoI is impacted and revision of the construct will be required.

Internal/External E-mail System

Email systems provide ability for users within an LMS to post messages that can be forwarded to external e-mail systems, whereby users are notified about new posted messages. While a mainstay of the LMS, email notifications are becoming passé among younger users. In place of email, users in this demographic prefer to rely upon texting, instant messenger clients, and so on. With the exception of lengthy communications, with a one-to-one focus, it is likely that we will soon see a convergence between these systems and Announcements tools, with the same implications noted above.

Chat

Chat is an online tool allowing for synchronous, communicative exchanges between/among multiple users within an LMS. In the early years of the LMS these tools were bulky and required numerous plugins to allow for one-to-one or group chat. Over the years, substantive advancements in these technologies have occurred, with virtual meeting clients (such as Connect and GoTo Meeting) being very robust and ubiquitous in academia. Notably, these same clients now have mobile applications, allowing for users to participate in chats and meetings virtually any time or anywhere. Likewise, simpler clients such as Skype and Facetime are all but ubiquitous and used regularly by a large percentage of students in their personal lives. Traditionally, these tools have been integrated into the LMS; however, as with many of the other tools discussed here, we must question why such integrations are necessary. From a technical perspective, such clients can be invoked by clicking on a simple link. Thus, the rationale for integration is more a function of simplifying faculty management of the learning environment as opposed to providing the most convenient pathway for students.

With respect to the impact on Social Presence, it was originally believed that these tools could have a positive impact on Group Cohesion as synchronous interaction was a novelty. However, now that these types of tools are in widespread use, we must consider how convenience and access will impact the same construct. While only anecdotal evidence exists and more research is needed, there is reason to believe that unless students are able to access chat/meeting tools without having to go through multiple access layers, there may be a resistance to using the system that will manifest in overall dissatisfaction with the entire Social Presence building process.

One final consideration that should be given to this class of tools is the potential impact of mature Virtual Learning Environments (VLEs). As previously noted, VLEs, such as Second Life, emerged with a great deal of hype, but disillusionment quickly set in as end user difficulties persisted. However, advancements continue to be made in this area, with the technology thresholds becoming more acceptable. If this trend continues, it is conceivable that VLEs could make significant inroads against more traditional chat/meeting tools. This in turn has the potential to strengthen Social Presence elements along the same lines that were envisioned when the construct was originally researched and developed.

Assignments/Activities/Course Content
Assignments is an important tool that serves both faculty and student needs. For instructors, the content editor allows the development of course assignments, activities, and content for the student to consume. Students also have access to a content editor within the discussion board tool, allowing for the reading, writing, and rendering of HTML text, images, and various multimedia enhancements.

This toolset lies at the heart of the LMS and the Cognitive Presence construct. As the place where course materials and activities are housed, this area is the focal point of courses. It is where core materials are provided and activities described. In the Cognitive Presence construct, it is where the instructor places problems that are intended to inspire students to engage in exploration and house materials for exploration. Notably, when the CoI was developed, text was the primary means for provisioning content. Despite having rich text editors, the content creation mechanisms within the LMS are rather clunky and limit faculty in terms of the content types that can be easily and quickly developed. While some third party providers have created content that can be integrated into the LMS, the overarching paradigm remains one of static content.

However, some advancements are being made in this area. For example, utilizing Adobe's Digital Publishing Suite (DPS), American Public University System (APUS) is undergoing an extensive restructuring of their online course offerings to include highly interactive components. Traditionally, DPS has been used by magazines such as *National Geographic, Wired,* and *Vanity Fair* to create rich, interactive digital experiences that can be consumed on the computer, tablet, or phone. The APUS has leveraged this technology to transform its classroom by including rich graphics, animations, videos, simulations, audios, and so on. Affordances such as touchscreen-tabbed interfaces, tooltips, and slideshows allow students to click or tap on a term or tab and receive more information, play embedded videos, receive information about points on a map, and engage with many other rich functionalities.[21]

From a theoretical perspective, these advances have the potential to significantly alter the Triggering Event and Exploration phases of the Cognitive Presence construct. Specifically, Triggering Event was conceptualized as a construct providing key questions or problems that would catalyze the desire to learn. While nuanced, the provisioning of experiences that have tactile engagement triggers accompanying problem posing scenarios may well change the way in which we think of creating advanced organizers for learning.

Likewise, while not specifically outlined as such, Exploration was envisioned as a series of discrete events in which students would engage with different types of learning assets as part of a pathway to learning. With the ability to aggregate various file types and fully leverage multimedia learning, as envisioned by Mayer and others, Exploration may need to be redefined to include elements that are dependent upon the media and states being utilized.

Grade Book

The Grade Book LMS tool allows for the storage, calculation, and distribution of student grades. The Grade Book is a vital tool that helps both instructors and students track progress and outcomes. Enrollment and group management are also traditionally facilitated through this toolset, with the enrollment aspect being largely automated and the group aspect being periodically adjusted by the instructor to meet short-term goals. In addition to its core functionality, the Grade Book provides a critical tie between institutions' student information systems and the LMS, through provisioning course rosters to the appropriate course shells. Course grades are then round-tripped back to the student information system, completing the administrative cycle.

While Grade Book tools are a lynchpin component, it should be noted that they are also available as stand-alone components, using such frameworks as JQuery to facilitate connections. Likewise, other stand-alone components can be tied to these types of Grade Book tools for purposes of transferring grades. Thus, it is possible to maintain this functionality without the need for a full LMS. At first glance, it would seem that this component would have little impact on elements of the CoI, as it is an administrative tool that plays little if any role in learning. However, the ability to maintain administrative functionality outside of the traditional

LMS framework allows for the formation of other tools, noted in this section, in a decentralized model.

Quiz/Testing Component

The quiz/testing component is an online tool providing opportunities to test their knowledge of course content. Quiz and testing options may include such formats as essay, multiple-choice, true/false, sequencing and fill-in-the-blank, and matching questions.

As with the Grade Book, this is an essential element of the LMS; however, there are component-based options. Additionally, there is momentum behind the idea of making assessments more authentic by embedding them within content or VLEs. At present, the Integration and Resolution phases of Cognitive Presence are envisioned largely as individual activities, resulting from collaboration in the Exploration phase. However, moving to more collaborative constructs, as made possible through contemporary technological affordances, would require significant rethinking of Cognitive Presence. Potentially this could result in constructs that relate to both personal and collaborative worlds.

Calendar

Instructors can use the calendar course tool to schedule, manage, and communicate events/assignments/due dates for course-related items. In early versions of the LMS, this functionality was considered extremely important as it was intended to provide a graphic representation of the due dates from the syllabus and keep students on track. From a theoretical perspective, it was also one of the foundational elements of the Instructional Design and Organization construct within Teaching Presence. However, this function has become marginalized through the ubiquity of calendar tools in email clients and on mobile devices. In fact, most LMSs now allow for administrators to push to such alternative tools and minimize or eliminate the calendar from the LMS interface.

CONCLUSION

Although seemingly inconsequential in terms of functionality, the Calendar tool is important to note because of the relative ease with which a major LMS component has been replaced over time. In a broader sense, it is representative of the disaggregation that is taking place within the LMS community in favor of an ecosystems approach. However, as practitioners, it is important that we look beyond the technological perspective and determine whether this also means that we are starting to see a disaggregation of the underlying learning theories, in particular the CoI.

While theorists and curriculum specialists have long held that changes in learning do not dictate pedagogy, it may be time for us to critically reassess this claim, especially in the realm of online learning. As described in the beginning of this chapter, the LMS provided the context against which the CoI was envisioned and defined. Maintaining a relatively steady state, in terms of innovation, or lack thereof, in the 15 plus years since their emergence, LMSs have also provided a consistent platform for validation and elaboration on the model. However, we are now approaching a nexus at which commercial innovation has evolved to the point where it is impinging on academic technologies, with the result that factures are occurring in the LMS-centric model.

We are not suggesting that the CoI is becoming invalid for describing online learning interactions. Rather, we suggest that the model needs to be rethought for various technological approaches, and a decision-tree type schema should be developed that accounts for variations and weighting difference for the three presences, depending upon which toolset and approach are being utilized. Perhaps the biggest challenge in accomplishing this will be in finding common ground between theorists and technologists to accompany this end. However, if we are to fully leverage emerging technological affordances, it is imperative that guiding theory be established to accompany such changes.

Notes

1. David C. Berliner, *The Development of Expertise in Pedagogy* (Washington, DC: AACTE Publications, 1988).
2. Eden Dahlstrom, Christopher Brooks, and Jacqueline Bichsel, *The Current Ecosystem of Learning Management Systems in Higher Education: Student, Faculty, and IT Perspectives* (Louisville, CO: ECAR, 2014).
3. Greg Kroner, "Understanding Small College Struggles," *Edutechnica*, accessed August 22, 2015, http://edutechnica.com/category/edtech/.
4. Eden Dahlstrom, J. D. Walker, and Charles Dziuban, *ECAR Study of Undergraduate Students and Information Technology* (Louisville, CO: ECAR, 2012).

5. Larry Cuban, *How Teachers Taught: Constancy and Change in American Classrooms, 1890–1990* (New York: Teachers College Press, 1993).

6. Ahmad Fawaz Alzaghoul, "The Implication of the Learning Theories on Implementing E-Learning Courses," *The Research Bulletin of Jordan ACM* 11, no. 11 (2012): 27–30.

7. Don W. Edgar, "Learning theories and historical events affecting instructional design in education: Recitation literacy towards extraction literacy practices," *Sage Open* 2 (2012): 1–9.

8. "Humanism," In *Learning Theories*, http://www.learning-theories.com/humanism.html.

9. "Twenty-First Century Skills," In *Learning Theories*, http://www.learning-theories.com/21st-century-skills-p21-and-others.html.

10. Brigham Fay, "Space Exploration: Today's Optimal Learning Environment is Adaptable, Transparent, and Connected in More Ways than One," *Usable Knowledge*. Harvard Graduate School of Education, February 17, 2015, http://www.gse.harvard.edu/news/uk/15/02/space-exploration.

11. Suzanne M. Wilson and Penelope L. Peterson, *Theories of Learning and Teaching: What Do They Mean*.

12. Cuban, *How Teachers Taught: Constancy and Change in American Classrooms, 1890–1990*.

13. John Dewey, *Dewey on Education. Selections with an Introduction and Notes by Martin S. Dworkin* (New York: Bureau of Publications, Teachers College, Columbia University, 1959).

14. Charlotte N. Gunawardena, "Social Presence Theory and Implications for Interaction and Collaborative Learning in Computer Conferences," *International Journal of Educational Telecommunications* 1, no. 2 (1995): 147–66.

15. See Anthony G. Picciano, "Beyond Student Perceptions: Issues of Interaction, Presence, and Performance in an Online Course," *Journal of Asynchronous Learning Networks* 6, no. 1 (2002): 21–40; Jennifer C. Richardson and Karen Swan, "Examining Social Presence in Online Courses in Relation to Students' Perceived Learning and Satisfaction," *Journal of Asynchronous Learning Networks* 7, no.1 (2003): 68–88; Karen Swan, "Building Learning Communities in Online Courses: The Importance of Interaction," *Education, Communication & Information* 2, no. 1 (2002): 23–49; Karen Swan, Randy Garrison, and Jennifer Richardson, *A Constructivist Approach*

to Online Learning: The Community of Inquiry Framework, in *Information Technology and Constructivism in Higher Education: Progressive Learning Frameworks*, edited by Carla R. Payne, 43–57 (Hershey, PA: IGI Global, 2009); Rupert Wegerif, "The Social Dimension of Asynchronous Learning Networks," *Journal of Asynchronous Learning Networks* 2, no. 1 (1998): 34–49.

16. Gunawardena, "Social Presence Theory and Implications for Interaction and Collaborative Learning in Computer Conferences."

17. Thomas F. Green, *The Activities of Learning* (New York: McGraw-Hill, 1971).

18. Paolo Friere, *Pedagogy of the Oppressed*, translated by Myra Bergman Ramos (New York: Herder and Herder, 1970).

19. Dee Hock, *Birth of the Chaordic Age* (San Francisco, CA: Berrett-Koehler Publishers, 1999).

20. Randy D. Garrison, Terry Anderson, and Walter Archer, "Critical Thinking, Cognitive Presence, and Computer Conferencing in Distance Education," *American Journal of Distance Education* 15, no. 1 (2001): 7–23.

21. See, e.g., https://itunes.apple.com/us/app/apus-lessons/id1015001464?mt=8.

BIBLIOGRAPHY

Alzaghoul, Ahmad Fawaz. 2012. The implication of the learning theories on implementing e-learning courses. *The Research Bulletin of Jordan ACM* 11(11): 27–30.

Berliner, David C. 1988. *The development of expertise in pedagogy*. Washington, DC: AACTE Publications.

"Constructivism." *Learning Theories*. Accessed August 1, 2015. http://www.learning-theories.com/constructivism.html.

Cuban, Larry. 1993. *How teachers taught: Constancy and change in American classrooms, 1890–1990*. New York: Teachers College Press.

Dahlstrom, Eden, J.D. Walker, and Charles Dziuban. 2012. *ECAR study of undergraduate students and information technology*. Louisville: ECAR.

Dahlstrom, Eden, D. Christopher Brooks, and Jacqueline Bichsel. 2014. *The current ecosystem of learning management systems in higher education: Student, faculty, and IT perspectives*. Louisville: ECAR. http://www.educause.edu/ecar.

Dewey, John. 1959. *Dewey on education. Selections with an introduction and notes by Martin S. Dworkin*. New York: Bureau of Publications, Teachers College, Columbia University.

Edgar, Don W. 2012. Learning theories and historical events affecting instructional design in education: Recitation literacy towards extraction literacy practices. *Sage Open* 2: 1–9.

Fay, Brigham. "Space Exploration: Today's Optimal Learning Environment is Adaptable, Transparent, and Connected in More Ways than One." *Usable knowledge*. Harvard Graduate School of Education. Accessed February 17, 2015. http://www.gse.harvard.edu/news/uk/15/02/space-exploration.

Friere, Paulo. 1970. *Pedagogy of the oppressed*. Trans. Myra Bergman Ramos. New York: Herder and Herder.

Garrison, D. Randy, Terry Anderson, and Walter Archer. 2001. Critical thinking, cognitive presence, and computer conferencing in distance education. *American Journal of Distance Education* 15(1): 7–23.

Green, Thomas F. 1971. *The activities of teaching*. New York: McGraw-Hill.

Gunawardena, Charlotte N. 1995. Social presence theory and implications for interaction and collaborative learning in computer conferences. *International Journal of Educational Telecommunications* 1(2): 147–166.

Hock, Dee. 1999. *Birth of the chaordic age*. San Francisco: Berrett-Koehler Publishers.

"Humanism." *Learning Theories*. Accessed August 1, 2015. http://www.learning-theories.com/humanism.html.

Kroner, Greg. Understanding small college struggles. Edutechnica. http://edutechnica.com/category/edtech/. Accessed 22 Aug 2015.

Linden Lab. Second life [software]. http://www.secondlife.com. Accessed 22 Aug 2015.

Picciano, Anthony G. 2002. Beyond student perceptions: Issues of interaction, presence, and performance in an online course. *Journal of Asynchronous Learning Networks* 6(1): 21–40.

Richardson, Jennifer C., and Karen Swan. 2003. Examining social presence in online courses in relation to students' perceived learning and satisfaction. *Journal of Asynchronous Learning Networks* 7(1): 68–88.

Swan, Karen. 2002. Building learning communities in online courses: The importance of interaction. *Education, Communication & Information* 2(1): 23–49.

Swan, Karen, Randy Garrison, and Jennifer Richardson. 2009. A constructivist approach to online learning: The community of inquiry framework. In *Information technology and constructivism in higher education: Progressive learning frameworks*, ed. Carla R. Payne, 43–57. Hershey: IGI Global.

"Twenty-First Century Skills." *Learning Theories*. Accessed August 1, 2015. http://www.learning-theories.com/21st-century-skills-p21-and-others.html.

Wegerif, Rupert. 1998. The social dimension of asynchronous learning networks. *Journal of Asynchronous Learning Networks* 2(1): 34–49.

Wilson, Suzanne M., and Penelope L. Peterson. 2006. *Theories of learning and teaching: What do they mean for educators?* Washington, DC: National Education Association.

CHAPTER 3

Designing Technology-Enhanced Active Learning Environments for the Undergraduate Geoscience Classroom

Priya Sharma and Kevin P. Furlong

Abstract This chapter reports on a multiyear collaborative research effort between geosciences and education, focusing on the design and evaluation of modules to engage undergraduate students in science reasoning skills. A design-based research approach was used to design active learning modules to engage students in a large general education undergraduate course in natural hazards, as well as to integrate mobile devices in an upper-level undergraduate course on the same topic. Over two iterations of each course, we were able to identify design features that supported student reasoning, as well as design of technology-enhanced learning within an undergraduate classroom supportive of collaborative student engagement in science reasoning.

Keywords Active learning • Design-based research • Geoscience education • Interdisciplinary research • Mobile devices • Science reasoning

Having a fundamental understanding of science and scientific reasoning is critical for all individuals so that they can engage in public discussions

P. Sharma (✉) • K.P. Furlong
Pennsylvania State University, University Park, PA, USA

© The Author(s) 2016
R.D. Lansiquot (ed.), *Technology, Theory, and Practice in Interdisciplinary STEM Programs*,
DOI 10.1057/978-1-137-56739-0_3

31

related to science and can be informed consumers of science and technology in their everyday life.[1] Science-reasoning skills are especially valuable to all students at the college/university level, not only for those majoring in science, technology, engineering, and mathematics (STEM) fields. A significant focus of science education, currently, is to engage students *actively* in the science reasoning and practices that are an inherent part of the scientific community,[2] rather than in the rote memorization of disconnected facts and concepts. In the geosciences, designing learning environments that most effectively support students is an important, yet nascent, area of focus.[3] This chapter focuses on a project that aimed to enhance science reasoning in undergraduate students (at both the general education and the upper-level science major levels) through an interdisciplinary teaching and research effort between geosciences and education. This interdisciplinary effort spanned approximately six years and two distinct foci of the project—one focused on a large-enrollment, technology-limited class, and the other focused on an upper-level, technology-rich class—are presented.

EDUCATING GEOSCIENTISTS: THEORY AND PRACTICE

Natural hazards have been proposed as a focal point for the Content Standard in *Science in Personal and Social Perspectives* for science education in K-12 as well as universities and societies.[4] The proposed national standard identifies the importance of linking the concept that "internal and external processes of the earth system cause natural hazards" to an understanding that "hazards can present personal and societal challenges because … incorrectly estimating the rate and scale of change may result in either too little attention and significant human costs or too much cost for unneeded preventive measures."[5] The focus of the National Science Education Standards (NSES) is on the importance of risk analysis, the physical processes that lead to natural hazards, and the important social decisions that must be made in regard to hazards. Although the NSES are focused at the K-12 environment, many of the goals and approaches of the NSES are equally valid and appropriate for STEM education in colleges and universities.

Although the goals associated with using natural hazards as a vehicle for science learning are relatively clear and straightforward, the realities of undergraduate education provide a series of hindrances to reaching those goals. At the general education level, science courses (like most general education courses) are often taught to very large enrollments in classrooms designed for lectures with limited opportunities for interactive and collaborative student work. Overcoming these infrastructural impediments requires a significant redesign of course content and delivery as well as a significant shift in pedagogical beliefs. At the upper level, the problems are different. Class size is typically reasonable (20–30 students) and facilities can be available for laboratory-based learning. However, the lab activities are too often disconnected from the course lecture, and there is little or no emphasis on science communication. Assessing the ability to clearly and accurately communicate science can be an effective measure of the overall success of the learning activity.

In this project, we have placed an emphasis on developing approaches to active learning that served the learning goals of the course and curriculum while recognizing the realities of classroom size and design, student backgrounds, and access to technology. The content of both courses used in this project is similar; both focus on natural hazards and their potential impact on society. The level of detail and scientific rigor varies with course level and student background, but the underlying scientific principles remain constant across both courses. Using a design-based approach,[6] we were able to implement two iterations of design in two different courses such that we are now able to comment on the effects of design iterations in individual courses on student achievement.

Phase 1: Integrating Active Learning in Large Undergraduate Courses

For the first phase of our project, we focused on developing active learning modules to engage students in lab-like active learning in the large enrollment environment of a lecture hall. The test bed for these modules was a 100-level general education course on natural disasters with an approximate enrollment of 160 or more students each semester. This context provided several interesting challenges for design, based primarily on the physical learning space, which was a large lecture hall, with fixed chairs and tables and a single instructor podium at the front of the lecture hall with a projector and white board.

To provide students the opportunity to conduct science in an authentic manner, the activities were designed using data sets and societal situations based on actual natural hazards and events. This design involved using a combination of larger science questions, such as determining possible temporal or spatial patterns to earthquake occurrence, and smaller questions, such as measuring the travel time for a tsunami from generation to hitting a coastline. In addition, we used common themes in the exercises that cut across subdisciplines and/or hazards. For example, the concept of energy was followed across numerous hazards, allowing students to compare the energy of a hurricane to the energy of an earthquake or that contained in a hot lava flow. We designed two modules (out of the approximately 15 total modules in the course) that were implemented similar to other modules used in the course; however, these modules had specific attributes that allowed for assessment of student learning. Every module in the course asks students to analyze a specific hazard/event and come to a conclusion or otherwise generate a science product. Embedded in each activity was an underlying science-reasoning goal, which fell into major themes and skills, including the ability to quantify patterns in space and time of natural events, and rates of processes; to couple scientific analyses to societal impacts; to incorporate real-time or near real-time data into the mix; and to compare the consequences of similar natural events in different environments (socioeconomic, cultural, political, etc.).

Design of Learning Modules for the First Iteration

The pedagogical framework for our design was active learning, which is generally defined as engaging students in meaningful learning activities as they think about what they are doing.[7] While this is a broad definition, most active learning designs include elements of collaborative/cooperative learning and problem-based learning.[8] For the initial iteration, active learning was implemented using three key design mechanisms: collaborative learning with authentic problems, scaffolding, and individual reflection. Peer collaboration has been shown to be effective in learning[9]; thus, students were engaged in several collaborative group activities as a means to encourage dialogue and discussion around geoscience concepts in the context of authentic natural disaster problems. In addition, two types of support were provided for students: *Procedural scaffolding*, which made explicit the sequence of activities for complex tasks and *cognitive scaffolding*, which helped learners reason through complex problems and guided them in "*what to consider.*"[10] Students were asked to consider several perspectives that dealt with real-life natural disaster problems and to respond to questions and provide reasoning for the decisions they made both as a group and individually.

The first active learning module was designed around a real-life complex problem related to hurricanes, titled "Hurricane Smith" and engaged students in decision-making processes for an evacuation plan in the event of an imminent hurricane. In this exercise, students were members of specified communities along the southeast US coast in Florida, South Carolina, or North Carolina; within each community, each student had a specific role and responsibility[11] such as being a member of emergency management groups or schools, and, as the hurricane developed, these groups would need to make decisions regarding evacuation and community management. The second active learning module, "Bangladesh Global Warming," used a similar structure and engaged students in a problem related to global warming, where students engaged in a small-group activity to identify regions most at risk from sea level rise.

After each group activity was completed, students were asked to write an individual report. Both individual reports required students to show clear links to background research and data and to provide reasoning for all their conclusions.

Implementation and Findings for the First Iteration

We evaluated the impact of the first iteration of the design by using pre- and post-tests related to the overall course content, as well as the individual and group artifacts produced by students in each of the two assessed activities, which were scored with a rubric developed specifically for such evaluation. The Geoscience Concept Inventory (GCI),[12] which was developed for assessing students' understanding of entry-level geoscience course material, was adapted for the pre- and post-test. To meet our goal of examining students' ability to employ scientific reasoning skills (e.g., quantitative reasoning, scientific thinking skills, geoscience concepts), additional items covering these areas were developed by the research team. Items were developed for three categories of thinking skills and concepts: scientific reasoning skills, quantitative reasoning skills, and geoscience concepts.

The pre-test was administered in the fourth session of the course, which was the second week (of a 15-week semester) of classes. Students were given 50 minutes to complete the pre-test. During the course, the Active Learning Modules I & II were implemented as part of the students' class activities during week 6 and week 12. Instructional materials for both active learning modules were distributed to all students, and they were randomly divided into groups of four or five for the in-class learning activities. The post-test was administered to the participants at the last session (i.e., 14 weeks after the pre-test) after all activities were completed. Students were given 40 minutes to complete the post-test.

To analyze the effect of active learning strategies on student learning, learning gains from pretest to post-test were compared with a paired t-test. Student learning was defined as changes in their performance on an assessment of quantitative reasoning skills, scientific reasoning skills, and geoscience concepts. The results of the paired t-test indicated a slight gain in mean pre- to post-test scores (10.25–10.28), with a p-value of 0.21, which was insignificant at the 0.05 level. To identify changes for specific sections of the test, follow-up paired t-tests were performed, which indicated that, on average, participants performed significantly higher on the geoscience concept section on the post-test ($M = 4.18$, SE = 0.15) than the pre-test ($M = 3.76$, SE = 0.18, $t(126) = -2.2$, $p < 0.05$, $r = 0.35$). However, students' performance in the quantitative reasoning section decreased from pre- to post-test with a p-value of 0.02, which was significant at the 0.05 level (see Tables 1 and 2).

Students' individual reports for the two modules were also examined using a paired t-test, and the results indicated that there was a gain in students' percentage mean scores from report I to report II, which was significant at the 0.05 level with a medium effect size (0.45) (see Tables 3 and 4).

These results prompted us to reconsider and redesign the modules for a second iteration of the study, as detailed below.

Table 1   Student performance means and standard deviation for the pre and post-test

Test	M	SD	Range	
			Low	High
Pretest[a]	10.25	3.75	2.0	17.5
Posttest	10.28	3.79	2.0	17.5

Note. Total score = 25

Source: Adapted from Kim et al. (2013)

[a] $n = 128$: participants for each test

Table 2 Student scores and means on inventory subsections

Subtest	Test	M	SD	t	df	Sig. (2-tailed)	Cohend
Geoscience concept section (n = 127)	Pretest	3.76	2.01	−2.15	126	(.03*)	0.22
	Posttest	4.17	1.73				
Quantitative reasoning section (n = 128)	Pretest	5.78	2.10	2.35	127	(.02*)	0.19
	Posttest	5.34	2.48				
Scientific reasoning section (n = 128)	Pretest	0.70	0.74	−.76	127	.45	0.08
	Posttest	0.76	0.72				
Total score (n = 128)	Pretest	10.25	3.75	−1.23	127	.21	0.01
	Posttest	10.28	3.79				

Note. Total score = 25. Subtest scores: Geoscience concept = 9, quantitative reasoning = 14, and scientific reasoning = 2

Source: Adapted from Kim et al. (2013)

*$p < 0.05$

Table 3 Percentage mean score and standard deviations for individual reports

Individual reports	Percentage mean score	SD	Range	
			Low	High
Individual reports I (HS)	68.34	19.36	16.7	100
Individual reports II (GW)	75.66	12.28	36.1	100

Note. Total score = 100

Source: Adapted from Kim et al. (2013)

$a°n$ = 105: participants for both reports

Table 4 Paired *t*-test results for individual reports

Individual reports	Mean difference	SD	t	df	Sig. (2-tailed)	cohend
I	7.31	21.20	3.53	104	(.00*)	0.45
II						

Note. Total score = 25. N = 105

Source: Adapted from Kim et al. (2013)

*$p < 0.05$

Second Iteration of Design for Active Learning in Large Undergraduate Courses

Based on the outcomes of the first-year study, we examined the data to identify design refinements for module design and data collection for the second iteration. The first design and implementation raised several issues. The main concern was the paucity of answers to the procedural question prompts as demonstrated in the group discussion charts for the first learning module (Hurricane Smith). We expected the directed prompts to help students to focus on the key issues they needed to deal with in making data-based decisions. Some groups omitted answers or did not provide details of their decision process or reasoning, while other groups provided very generic responses that did not indicate that students used the resources made available for the hurricane module. Another hindrance was the use of numerous discussion sessions. Students met for discussion six times (within a 50-minute class period) with their classmates who had the same roles, and the six iterations were deemed an ineffective use of time. Based on this re-analysis of the first year's implementation, the main modification for the second year was made in iteratively refining two design principles: peer interaction and question prompts.

Changes for Peer Interaction: In iteration 1, students were engaged in too many different processes to make decisions appropriately, and the worksheets indicated an insufficient understanding of hurricanes as well as individual student roles within the given community. The following design changes were indicated:

- A need for more individual preparation in order to improve peer interaction activities;
- Reducing the steps for community discussion making (two community discussions, instead of three community discussions); and
- Asking students to focus more clearly on important aspects of the problem based on their roles, and eliminating the role-based group work discussion.

Changes for question prompts: Students failed to show concrete reasoning for their decision-making regarding community evacuation procedures during the hurricane. The following design changes emerged:

- A need for procedural question prompts to support group decision-making; and

- A need for elaborative question prompts in the process of individual preparation and group discussions.

Implementation and Results for the Second Iteration

The revised modules were implemented in a subsequent semester of the course and group and individual artifacts were collected after each module was finished. For the second iteration, we focused primarily on individual critical thinking over the two iterations of the modules (i.e., for year 1 and year 2). In terms of total scores of critical thinking, the students in the second year ($M = 25.45$, SD = 6.311) scored significantly higher than those of the first year ($M = 23.32$, SD = 7.742), [$t(129) = -2.432$, $p < 0.05$]. The lower 25 % students of the second year ($M = 17.81$, SD = 2.978) also scored significantly better than those of the first year ($M = 14.19$, SD = 3.316), [$t(31) = -4.601$, $p < 0.01$]. However, there was no significant difference between the top 25 % students of first and second years [$t(31) = 1.041$, $p > 0.05$]. Overall, the students of the second year demonstrated higher levels of critical thinking and the difference was especially evident in the lower 25 % of students (see Table 5). These findings imply that the second year design, which employed procedural and elaborative question prompts and peer interaction with individual preparation, was more effective in enhancing students' critical thinking and worked better for the lower-level than the top-level students.

While the numerical data indicated that the second iteration of design was more successful in increasing student performance in the individual critical thinking reports, we also wanted to explore whether there were any

Table 5 Independent t-test in critical thinking between first and second years

	First year			Second year			t	Sig	Cohend
	n	M	SD	n	M	SD			
Top 25 %	32	34.38	1.792	32	33.94	1.564	1.041	0.302	0.261
Lower 25 %	32	14.19	3.316	32	17.81	2.978	−4.601	0.000[a]**	1.149
Total	130	23.32	7.742	130	25.45	6.311	−2.432	0.016[b]*	0.302

Source: Adapted from Yoo (2011)

[a]$p < 0.01$ level; [b]$p < 0.05$ level

changes within the group decision-making. The data gathered from group decision charts were compared across the two years, revealing that groups in the second iteration addressed almost all of the decision points during the group activities, which directly addressed the design change to reduce and emphasize the decision points in the second iteration. Moreover, they provided more concrete and elaborate reasoning and data within their recommendations, which directly addressed the design change to include elaborative and procedural prompts to support group reasoning. We also used a multiple case study design[13] to explore the links between the group-generated documents and the individual reports, to identify if there were any possible explanations that could link learning activities with individual critical thinking. Using a pattern match technique, data were organized to support plausible explanations about the relationship between the learning activities and critical thinking. The main approach to data analysis involved using theoretical propositions, developing a case description, and using both qualitative and quantitative data. Especially in the context of using role-play, these analyses indicated that higher-scoring students focused more on their assigned role, and specifically linked their roles to decisions they made. In addition, students who provided detailed answers within the individual worksheets and provided appropriate and detailed data tended to be higher-scoring students.

Phase 2: Technology-Enhanced Learning in Upper-Level Geoscience Courses

As a complement to the project described above that was aimed at improving the effectiveness of science learning through the use of hands-on, active learning modules within the large-enrollment general education classroom, we have also worked on enhancing an upper-level natural hazards course by integrating technology into specific learning activities. Our goal in this project was to let the technology allow the students to better replicate the science analysis experience, rather than to use technology solely to deliver content. That is, we have attempted to incorporate the technology as a method to increase the authenticity of the activity.

Different sciences approach research in different ways, ranging from substantially observational to purely theoretical. But underneath these superficial differences, there are similarities to most scientific investigations that are articulated in the following list of practices for science[14] such that educators can focus attention on supporting and making these practices explicit:

1. Asking questions
2. Developing and using models
3. Planning and carrying out investigations
4. Analyzing and interpreting data
5. Using mathematics and computational thinking
6. Constructing explanations (for science) and designing solutions (for engineering)
7. Engaging in argument from evidence
8. Obtaining, evaluating, and communicating information

The upper-level course that is the focus of the second phase of our collaborative research efforts offers various opportunities to engage students in socioscientific issues that help integrate scientific arguments and claims with political and ethical choices about actions.[15] For example, it is natural to combine computations and assessments of natural hazards with societal and cultural impacts. Indeed, examining and making decisions about socioscientific issues is a key goal for science education. Our goal in this phase of the project was to enhance students' active engagement in the content and problems by seeking out and working with authentic data and sources, and working with peers by leveraging social media, mobile devices, and open content.

First Iteration of Design for Technology-Enhanced Geoscience Modules

This project focused on the learning environment in a specific upper-level science course, Natural Disasters. The course is aimed at a diverse group of upper-level undergraduate students and is typically populated with students from a wide range of majors including Geosciences, Geography, Meteorology, Civil Engineering, Education, and Energy Business and Finance. Students often take this course as one of their final courses in their undergraduate program, and so there is a focus on improving their higher-level science reasoning and helping them articulate that reasoning.

The course content is built around a series of topical case studies, oftentimes using recent disasters as the basis for the analyses. Whether it is about the deadly earthquakes in Haiti, Japan, or New Zealand, or the effects of Hurricane Ike, the common theme to the course is having students analyze the events scientifically and then apply those results and improved understanding of the causes and consequences of the event to a specific task. For example, after analyzing the effects of specific hurricanes in the USA, the students may be asked to analyze what would happen if a similar event occurred elsewhere—such as in Bangladesh, and write a 'white paper' report to the relevant UN agency. In this way the students are encouraged to transfer their science-content knowledge to a new situation, assess the consequences, and communicate their reasoning. This course has traditionally been offered as an on-campus course, at least once (and oftentimes twice) each year. As a writing intensive course, its enrollment is limited to 25 students.

For the past few semesters, students have used iPads to access online content and as a tool for in-class activities, this addressing several fundamental affordances: one, providing students with access to primary data from key sources (e.g., observations and other data hosted by science-focused government agencies such as the USGS, NOAA, and NASA), and two, providing students with apps and tools to appropriately use and integrate the data. Although the data are easily accessible, effective use of those resources requires appropriate tools to use them for authentic scientific practices.

All the iPads had the same set of apps installed. In addition to standard apps such as web browsers, the specific apps we used ranged from broad use tools (*GoodReader, Wolfram Alpha, iWorks Suite*) to those specific to activities, data resources, and learning objectives in the geosciences (*Epicentral+, Elevation Chart, GeoMeasure*), and some general purpose quantitative and modeling tools that are particularly valuable to geoscience analyses (*Stella Modeler, Elevation Chart, TopoMaps. GeoMeasure*).

Overall, the purpose of using iPads and apps is to provide a means for students to move easily and efficiently from (a) asking a question to (b) getting relevant data to (c) exploring, modeling, and elaborating the issues raised by their scientific enquiry. In this context, iPads expand the information space available to students by not only providing access to content, but also introducing experts who can augment their understanding or practice of specific disciplines. In this case, expert content is represented by the data and resources made available through the USGS and NOAA. Such activities also map well to the top five uses of mobile devices as identified by the ECAR study[16] and are generally in line with the expectations of students in using familiar technology and practices to support their learning in the classroom and out.

Implementation and Findings for Technology-Enhanced Geoscience Modules

In spring 2014, we collected data on students' use of iPads within the course, and we used an ethnographic framework[17] to document their work with the iPads and other materials *in situ*. This initial data collection was treated as a way for the research team to make some initial assessments of student use of iPads and apps to gain more understanding about the following aspects: (a) how the instructional goal and activities affected student use of the iPads; (b) whether students collaborated (or not) with each other while using the iPads; (c) how students specifically used affordances of the iPads to support reasoning about tasks; and, (d) how students used resources other than iPads and apps to address instructional tasks.

Since our goal was to examine and understand student use of iPads in context, we used video-based fieldwork[18] to capture video and audio records of class sessions where the students used iPads. We recorded eight class sessions over three months, and positioned multiple cameras and audio mics to capture as much detail as possible. While data analyses are still ongoing, preliminary data analyses indicate the following key design foci:

- *More devices, less interaction*: Approximately 12 iPads were available to the class, which had 23 students. Thus, in an average class session, every pair shared an iPad. In some cases, students brought their own laptops or used their personal mobile devices, and it was noted that the more devices on the table (i.e., the closer the group got to a 1:1 ratio) the less interaction and discussion among students. Instead of collaborating with peers, students chose to instead work individually and occasionally share findings.
- *Navigating multiple resources requires support*: Given the nature of the course, students often had multiple media at their tables, including paper maps, rulers, iPads, worksheets, and so on. Especially for measurement tasks, such as calculating distances and slopes, and so on, students seemed to use the paper maps and the *TopoMaps* app on the iPad in different ways to support their tasks. In some instances, even though the *TopoMap* app had capabilities not present in the

papers maps, students chose to use the paper maps, perhaps out of habit. Also students often used their mobile phones to support task activities, such as taking pictures of graphs or worksheets for future reference.

- *Instructors focused on technical support vs. scaffolding*: Instructors often spent time to help groups with the iPad interface and apps, and often the same type of help was offered multiple times to different groups in a class session. However, the instruction was aimed more at routine use of the interface (e.g., how to measure distances or how to calculate slope) rather than more valuable scaffolding to support student reasoning about the task and ways to navigate the multiple resources and interfaces available.

- *Student facility with devices differed*: While not surprising, it is worth noting that some students were more comfortable using and exploring the iPad apps than others. The difference in using mobile phones as opposed to iPads was quite notable. Every student in the class had a personal mobile phone and was seen using it to search the web, send text messages, and take pictures. However, the facility with mobile phones was not demonstrated with or transferred to the use of iPads and indeed many seemed to have trouble identifying which apps to use or how to connect to the Internet.

Second Iteration of Technology-Enhanced Geoscience Modules

Based on the findings of the previous iteration, we decided to make the following design changes to the next iteration. First, we changed the physical location of the classroom and moved the class sessions to a lab that supported flexible learning arrangements, both in terms of how students could be seated, as well as how groups could use multiple display screens to project their interim findings. This change was occasioned by other research, which has indicated that having shared displays has a positive impact on students' participation, social interaction, and collaboration.[19] To focus our attention more clearly on the group's tasks and interaction, we also decided to draw on our previous projects' findings and design problem-based activities with elaborative and conceptual scaffolds that required the groups to find authentic data, manipulate that data, and negotiate the findings as a group. Also, since this was an upper-level class, instead of providing detailed procedural scaffolding, as we had for the previous iteration, we decided to provide conceptual scaffolds on how to address the problem. We also focused our problem design more clearly around socioscientific issues to clearly articulate the connection between science and its societal impacts.[20]

Implementation and Findings for the Second Iteration

In this second iteration, we adapted the modules to more systematically build on previous materials over a 3–4 class period duration. This allowed students ample time to consolidate materials and do some independent background research as the activity proceeded. An example of this linkage and scaffolding is the tsunami activity. We began with an analysis of the tsunami that devastated Japan in 2011, using observations of tsunami heights and travel times. With this introduction, students were able to (1) develop skills in using the various apps and equations related to tsunami propagation, (2) calibrate their analyses against observations, and (3) see the consequences of such a tsunami on the region. The students then moved to undertake similar analyses along the coast of Oregon in the USA, a region at risk of a similar tsunami as experienced in Japan. Finally, once the physical characteristics of a tsunami in Oregon were understood, the students were asked to evaluate an evacuation plan for Cannon Beach, a seaside resort town in Oregon. Through this systematic application of the science of tsunamis and its societal impact, we intended that the students would not only see the value of such analyses in a retrospective mode, but also develop an understanding of the predictive power of well-calibrated scientific analyses. Throughout the module, access to apps on the iPads provided a means for students to quantitatively address the primary questions of tsunami speed, height, and timing.

To gather data on how students engaged with the technology and each other, we video-recorded all sessions where the students used the iPads. We also collected screen-capture data for 1 to 4 iPads each session. The videos of the student work were then synced with the recorded iPad activity and imported into Studiocode, video analysis software. Combining these different perspectives allowed us to look at students' actual activity on the iPad and coordinate their discourse and interactions with their group members to understand how the different technologies were being appropriated to support their problem-solving and task processes. We also implemented a pre- and post-survey to help us gauge students' familiarity with the different technologies before and after the course. We implemented the pre-survey in the fourth week of the course, before the designed modules were implemented; the post-survey was implemented in the last week of the course. While data analyses for the video and iPad data are still ongoing, we were able to identify the following results from the surveys.

Seventeen students completed the pre- and post-survey, which asked a variety of questions related to their familiarity with different devices and apps, as well as their prior knowledge of using apps. In the pre-survey, only three students reported having any prior experience using iPads in the classroom, although almost 90% of them reported having both a laptop computer and mobile phone, and 40% reported having a tablet device. Students also reported being more comfortable with using laptops for academic work. Based on the post-survey data, students' self-reported self-efficacy in using *TopoMap*, *Keynote*, *Pages*, *ElevationChart* and *Wolfram Alpha* was significantly improved after taking the course ($p = 0.000$, 0.009, 0.004, 0.000, and 0.006). Students also had a generally positive impression of iPads in the classroom: the majority of participants strongly agreed that iPads supported their groups in addressing the tasks that they were actively involved in the course when they used iPads, and that the iPads helped them to engage with the activities.

While analyses for the group video and audio captured data are still ongoing, we envision the results of the analyses allowing us to iteratively refine the design modules that will be implemented in the Natural Disasters course in future semesters, as well as allowing us to better understand how to improve student learning within a technology-enhanced geoscience course.

CONCLUSION

This project sought to investigate potential for the long-term engagement of education and geosciences collaborators via both iterations of the project. The major implications and conclusions arising from this long-term interdisciplinary research project can be encapsulated in the following themes: long-term engagement in projects; negotiating pedagogical and conceptual variations in disciplines; and, effective incorporation of technology in science teaching. Using a design-based approach for both projects was significantly helpful in making both conceptual (such as in the case of tweaking the group discussion and individual worksheets for the first project) and concrete (such as changing the physical affordances of the classroom) adjustments to the design, such that modules could better support learning. The long-term engagement also helped collaborators in both disciplines to gain more familiarity with the conceptual tools and processes inherent in each discipline; as an example, in creating the modules for the upper level technology-enhanced course, the education team would have preferred to have individual students engage in more sustained research and data gathering before embarking on group work. However, as pointed out by the geoscience team, most geoscientists working in agencies such as USGS and NOAA are trained to make quick assessments in limited time frames based on currently available and historical data, especially as natural hazard events often occur at unpredictable times (e.g., earthquakes and tsunamis). Thus, in the final iteration of the modules, our designs tried to incorporate both pedagogically relevant principles while accounting for the real time, practical constraints faced by professional geoscientists.

This example also ties into the second theme of negotiating conceptual and pedagogical variations in disciplines. Different disciplines approach teaching differently[21] and certain pedagogical modes are more prevalent in some disciplines than in others. As an example, natural science teaching is likely more dominated by lab teaching, field trips, and exercises, while humanities are dominated more by seminars, workshops, and tutorials.[22] We did not encounter such a divergence of pedagogy in this project. For example, both teams were highly in favor of active, engaged, student learning in the classroom as well as the use of collaborative and authentic technology-enhanced learning to support student learning. However, the teams differed slightly on the emphasis of design. For instance, in the initial iteration of the technology-enhanced classroom design, the geoscience team favored a more open-ended approach to integrating technology (i.e., in providing the mobile devices and resources and allowing students to freely explore the apps), whereas the education team favored a more staged approach to integrating technology, and providing technology and task training to the students. The results of our first study in that context suggested that a blended approach would be best, where technology and practices were introduced within the context of the learning activity with some exploratory affordances to allow students varying abilities with technology to assist in the process as needed.

From a purely pedagogical level, our design of the intervention in both phases of the project was largely determined by the context and the audience that we were addressing. For example, it was more challenging to implement technology in a meaningful way in the large enrollment course as opposed to the smaller, upper-level course. The upper-level course allowed us to investigate the impact of implementing technology in the learning environment—both at the classroom scale and at the individual student level. Our initial results from that (still ongoing) phase of the project can be summarized as follows. First, although technology, and in particular smart devices (phones, tablets, etc.) are quite familiar to all of the students, their use within a scientific mode is new to many. As a result, the effective use of specific apps and even the approach to using such devices for science must be explicitly included in the preparatory materials. Second, we saw a major change in student engagement in moving to a classroom that easily allowed students to share materials and results. Both classrooms used allowed students to meet in groups around a table or cluster of tables, but the second-year classroom that easily allowed the technology to be shared both within each group and with the overall class had (to our observations) substantially more student engagement. Third, as the number of smart devices available per student approached a 1:1 ratio, the amount of student interaction decreased significantly. A 1:3 or 1:2 ratio was far more effective in fostering student interaction.

Consequently, our engagement in this long-term, collaborative design and research project allowed us to iteratively refine design modules to meet different contexts and audiences, albeit with somewhat similar learning goals. The outcomes of our first project indicated that active learning can better support students to engage in scientific reasoning in large enrollment classes, and, through our second project, we anticipate being able to show how designed integration of technology can support students' engagement in scientific practices. We anticipate that both geoscience and education can benefit from these studies, and that we can continue this interdisciplinary endeavor as a mechanism to contribute further to design and pedagogy research in geoscience and education.

NOTES

1. National Research Council, *A Framework for K-12 Science Education: Practices, Crosscutting Concepts, and Core Ideas,* eds. Helen Quinn, Heidi Schweingruber, and Thomas Keller (Washington, DC: National Academies Press, 2012); NGSS Lead States, "*Next Generation Science Standards: For States, by States,* (Washington, DC: National Academies Press, 2013).
2. Tim J Parkinson et al., *Engaging Learners Effectively in Science, Technology and Engineering: The Pathway from Secondary to University*

Education (Wellington, New Zealand, AOTOERA, 2011) https://akoaotearoa.ac.nz/download/ng/file/group-1657/engaging-learners-effectively-in-science-technology-and-engineering---full-report.pdf; National Research Council, *National Science Education Standards, Science Education* (National Academy Press, 1996).

3. Chris King, "Geoscience Education: An Overview," *Studies in Science Education* 44, no 2 (2008): 187–222.

4. National Research Council, *National Science Education Standards* (Washington, DC: National Academies Press, 1996).

5. Ibid., 168.

6. The Design-based Research Collective, "Design-Based Research: An Emerging Paradigm for Educational Inquiry," *Educational Researcher* 32, no. 1 (2003): 5–8, doi:10.3102/0013189X032001005; Sasha Barab and Kurt Squire, "Design-Based Research: Putting a Stake in the Ground," *Journal of the Learning Sciences*, 2004.

7. Michael Prince, "Does Active Learning Work? A Review of the Research," *Journal of Engineering Education* 93, no. July (2004): 223–31.

8. Ibid.

9. Norris Armstrong, Shu Mei Chang, and Marguerite Brickman, "Cooperative Learning in Industrial-Sized Biology Classes," *CBE Life Sciences Education* 6, no. 2 (2007): 163–71; Timothy T Eaton, "Engaging Students and Evaluating Learning Progress Using Collaborative Exams in Introductory Courses," *Journal of Geoscience Education* 57, no. 2 (2009): 113–12; Marlene Scardamalia and Carl Bereiter, "Adaptation and Understanding: A Case for New Cultures of Schooling," in *International Perspectives on the Design of Technology-Supported Learning Environments*, ed. Stella Vosniadou et al. (New York: Lawrence Erlbaum Associates, 1996), 149–63.

10. Michael Hannafin, Susan Land, and Kevin Oliver, "Open Learning Environments: Foundation, Methods, and Models," In *Instructional-Design Theories and Models: A New Paradigm of Instructional Theory, Vol. II.* ed. Charles M. Reigeluth, 114–40 (Mahway, NJ: Lawrence Erlbaum, 1999).

11. Kristina M. DeNeve and Mary J. Heppner, "Role Play Simulations: The Assessment of an Active Learning Technique and Comparisons with Traditional Lectures," *Innovative Higher Education*, 1997; Bram De Wever et al., "Roles as a Structuring Tool in Online Discussion Groups: The Differential Impact of Different Roles on

Social Knowledge Construction," *Computers in Human Behavior* 26, no. 4 (2010): 516–23.

12. Julie C. Libarkin and Steven W. Anderson, "Assessment of Learning in Entry-Level Geoscience Courses: Results from the Geoscience Concept Inventory," *Journal of Geoscience Education* 53, no. 4 (2005): 394–401.

13. Robert K. Yin, *Case Study Research: Design and Methods*, Applied *Social Research Methods Series*, 5th edition (Thousand Oaks, CA: Sage, 2009).

14. National Research Council, *A Framework for K-12 Science Education: Practices, Crosscutting Concepts, and Core Ideas.*

15. Stein Dankert Kolstø et al., "Science Students' Critical Examination of Scientific Information Related to Socioscientific Issues," *Science Education* 90, no. 4 (2006): 632–55.

16. Eden Dahlstrom, J.D. Walker, and Charles Dziuban, "ECAR Study of Undergraduate Students and Information Technology, 2013," *Educause Centre for Applied Research*, (2013): 1–12, https://net. educause.edu/ir/library/pdf/ERS1302/ERS1302.pdf.

17. James P Spradley, "Participant Observation," ed. B Crabtree and W Miller, *Qualitative Research*, Applied social research methods series TS—Österreichischer Verbundkatalog 3, no. 2 (1980): 1–16, http:// qrj.sagepub.com/cgi/doi/10.1177/14687941030032003\ nhttp://www.citeulike.org/group/8921/article/4046568.

18. Sharon J Derry, (Ed.), *Guidelines for Video Research in Education: Recommendations from an Expert Panel* (Chicago, IL: National Science Foundation, Data Research and Development Center, 2007), accessed August 20, 2015, http://drdc.uchicago.edu/ what/video-research-guidelines.pdf.

19. Chen Chung Liu et al., "Analysis of Peer Interaction in Learning Activities with Personal Handhelds and Shared Displays," *Educational Technology and Society* 12, no. 3 (2009): 127–42.

20. Troy D. Sadler and Dana L. Zeidler, "Scientific Literacy, PISA, and Socioscientific Discourse: Assessment for Progressive Aims of Science Education," *Journal of Research in Science Teaching*, 46 (2009): 909–21.

21. George R Lueddeke, "Professionalising Teaching Practice in Higher Education: A Study of Disciplinary Variation and 'Teaching-Scholarship,'" *Studies in Higher Education* 28, no. 2 (2003): 213–28.

22. Ruth Neumann, "Disciplinary Differences and University Teaching [Electronic Version]," *Studies in Higher Education* 26, no. 2 (2001): 135–46, doi:10.1080/0307507012005207.

BIBLIOGRAPHY

Armstrong, Norris, Shu Mei Chang, and Marguerite Brickman. 2007. Cooperative learning in industrial-sized biology classes. *CBE Life Sciences Education* 6(2): 163–171.

Barab, Sasha, and Kurt Squire. 2004. Design-based research: Putting a stake in the ground. *Journal of the Learning Sciences* 13(4): 1–14.

Dahlstrom, Eden, J. D. Walker, and Charles Dziuban. 2013. ECAR study of undergraduate students and technology, 2013. *Educause Centre for Applied Research*: 1–12. https://net.educause.edu/ir/library/pdf/ERS1302/ERS1302.pdf.

De Wever, Bram, Hilde Van Keer, Tammy Schellens, and Martin Valcke. 2010. Roles as a structuring tool in online discussion groups: The differential impact of different roles on social knowledge construction. *Computers in Human Behavior* 26(4): 516–523.

DeNeve, Kristina M., and Mary J. Heppner. 1997. Role play simulations: The assessment of an active learning technique and comparisons with traditional lectures. *Innovative Higher Education* 21(3): 231–246.

Derry, Sharon J., ed. 2007. *Guidelines for video research in education: Recommendations from an expert panel*. Chicago: National Science Foundation, Data Research and Development Center. Accessed August 20, 2015. http://drdc.uchicago.edu/what/video-research-guidelines.pdf.

Eaton, Timothy T. 2009. Engaging students and evaluating learning progress using collaborative exams in introductory courses. *Journal of Geoscience Education* 57(2): 113–120.

Hannafin, Michael, Susan Land, and Kevin Oliver. 1999. Open learning environments: Foundation, methods, and models. In *Instructional-design theories and models: A new paradigm of instructional theory*, vol. II, ed. Charles M. Reigeluth, 115–140. Mahwah: Lawrence Elrbaum.

Kim, Kyoungna, Priya Sharma, Susan M. Land, and Kevin P. Furlong. 2013. Effects of active learning on enhancing student critical thinking in an undergraduate general science course. *Innovative Higher Education* 38(3): 223–235.

King, Chris. 2008. Geoscience education: An overview. *Studies in Science Education* 44(2): 187–222.

Kolstø, Stein Dankert, Berit Bungum, Erik Arnesen, Anders Isnes, Terje Kristensen, Ketil Mathiassen, Idar Mestad, Andreas Quale, A.S.V. Tonning, and Marit Ulvik. 2006. Science students' critical examination of scientific information related to socioscientific issues. *Science Education* 90(4): 632–655.

Libarkin, Julie C., and Steven W. Anderson. 2005. Assessment of learning in entry-level geoscience courses: Results from the geoscience concept inventory. *Journal of Geoscience Education* 53(4): 394–401.

Liu, Chen Chung, Chen Wei Chung, Nian Shing Chen, and Baw Jhiune Liu. 2009. Analysis of peer interaction in learning activities with personal handhelds and shared displays. *Educational Technology and Society* 12(3): 127–142.

Lueddeke, George R. 2003. Professionalising teaching practice in higher education: A study of disciplinary variation and 'teaching-scholarship'. *Studies in Higher Education* 28(2): 213–228.

National Research Council. 1996. *National science education standards. Science education*. Washington, DC: National Academies Press.

National Research Council. 2012. In *A framework for K-12 science education: Practices, crosscutting concepts, and core ideas*, eds. Helen Quinn, Heidi Schweingruber, and Thomas Keller. Washington, DC: National Academies Press. http://www.nap.edu/catalog.php?record_id=13165. Accessed 20 Aug 2015.

Neumann, Ruth. 2001. Disciplinary differences and university teaching [electronic version]. *Studies in Higher Education* 26(2): 135–146. doi:10.1080/0307507012005207.

NGSS Lead States. 2013. *Next generation science standards: For states, by states.* Washington, DC: National Academies Press.

Parkinson, Tim J., Helen Hughes, Dianne H. Gardner, Gordon T. Suddaby, Marg Gilling, and Bill R. MacIntyre. 2011. *Engaging learners effectively in science, technology and engineering: The pathway from secondary to university education.* Wellington: AOTOERA. Accessed August 20, 2015. https://akoaotearoa.ac. nz/download/ng/file/group-1657/engaging-learners-effectively-in-science-technology-and-engineering---full-report.pdf.

Prince, Michael. 2004. Does active learning work? A review of the research. *Journal of Engineering Education* 93(3): 223–231.

Sadler, Troy D., and Dana L. Zeidler. 2009. Scientific literacy, PISA, and socioscientific discourse: Assessment for progressive aims of science education. *Journal of Research in Science Teaching* 46: 909–921.

Scardamalia, Marlene, and Carl Bereiter. 1996. Adaptation and understanding: A case for new cultures of schooling. In *International perspectives on the design of technology-supported learning environments*, ed. Stella Vosniadou, Erik De Corte, Robert Glaser, and Heinz Mandl, 149–163. New York: Lawrence Erlbaum Associates.

Spradley, James P. 1980. "Participant observation," edited by B Crabtree and W Miller. *Qualitative Research, Applied Social Research Methods Series TS— Österreichischer Verbundkatalog* 3(2): 1–16.

The Design-based Research Collective. 2003. Design-based research: An emerging paradigm for educational inquiry. *Educational Researcher* 32(1): 5–8. doi: 10.3102/0013189X032001005.

Yin, Robert K. 2009. *Case study research: Design and methods*, Applied social research methods series, 5th ed. Thousand Oaks: Sage.

Yoo, Suhyun. 2011. Enhancing students' critical thinking in science: A two-year design based exploration in a large undergraduate science course. (Doctoral dissertation). Pennsylvania State University.

CHAPTER 4

Educating Students for STEM Literacy: GlobalEd 2

Kimberly A. Lawless, Scott W. Brown, and Mark A. Boyer

Abstract GlobalEd 2 (GE2) engages classrooms of students online, and simulates negotiations of international agreements on issues of global concern such as water scarcity and climate change. GE2 is an interdisciplinary problem-based curriculum targeting students' global awareness, scientific literacies, and twenty-first century workforce skills. For the past 15 years, various iterations of GE2 have been implemented in classrooms, ranging from middle schools through college. Results have demonstrated the positive impact of GE2 along a number of dimensions including writing, argumentation, science knowledge, and social perspective taking. This chapter provides an overview of GE2, its design principles and discusses data from a recent implementation with college freshmen, specifically focusing on gains with respect to self-efficacy across multiple domains.

Keywords Global education • Problem-based learning • Science education • Simulation • STEM education • Writing skills

K.A. Lawless
College of Education, University of Illinois at Chicago, Chicago, IL, USA

S.W. Brown (✉)
Educational Psychology, University of Connecticut, Storrs, CT, USA

M.A. Boyer
Environmental Studies Program, University of Connecticut, Storrs, CT, USA

© The Author(s) 2016
R.D. Lansiquot (ed.), *Technology, Theory, and Practice in Interdisciplinary STEM Programs*,
DOI 10.1057/978-1-137-56739-0_4

53

Today's college graduates are entering an interconnected world in which globalization and science, technology, engineering, and mathematics (STEM) literacy will affect nearly every facet of their lives. Yet, as current and future citizens, our students' global awareness and STEM literacy remain strikingly limited—if anything, they appear to be in a state of decline. According to Derek Bok, the USA bears "the dubious distinction of being one of only two countries in which young adults were less informed about world affairs than their fellow citizens from older age groups."[1] Compounding the problem is the complexity of global learning itself: a balance of knowledge (such as science, geography, politics, and economics), skills (the ability to find and evaluate information sources and communicate their meaning to others), and attitudes (interest, efficacy, appreciating the value of other cultures, and having a sense of responsibility for our shared planet). Helping students to acquire such a diverse array of knowledge, skills, and attitudes cannot be accomplished through a single discipline. Nor can it be taught with traditional expectations of disciplinary mastery, since its subject is constantly changing and is as vast as the globe itself.

Concomitant with the need to develop global citizens, colleges are also responsible for preparing students for the twenty-first century work force, which is also rapidly evolving. Across several independent surveys of businesses and potential employers, the most commonly cited skills that industries require in newly graduated college students include the following: the abilities to solve complex, multidisciplinary problems; to work successfully in teams; effective oral and written communication skills; and good interpersonal skills. Yet in report after report, employers reported that universities are failing to prepare graduates for the current expectations of the workforce.

Preparing students for the globalized world and the twenty-first century workforce requires that both application and relevance be present. John Morley suggested that students should "know how rather than simply knowing that,"[2] and John Heldrich stated in 2005, "higher education can be improved by making it more relevant to what happens in the workforce."[3] According to Derek Bok, this is best accomplished through intentional educational practices that are integrative in nature, provide experiences that challenge students' own embedded worldviews, encourage application of knowledge to contemporary problems, and integrate knowledge across a wide array of disciplines. Creating and implementing educational experiences with these characteristics in higher education will not only develop critical thinking skills among our students, but will also equip them as citizens with the drive, values, capacity to question, and ability to develop solutions that will help them advance both commercial and social interests.

While it is clear that these preparation gaps of today's college students exist and have existed for some time, most universities have made little progress toward resolving these deficits. Institutions of higher education continue to operate using programmatic approaches that exacerbate the siloed, decontextualized nature of academic content and skill sets—an approach that is counter-productive to facilitating twenty-first century skills. The genesis of this emanates from a number of structural and financial issues regarding how universities operate. Universities are organized in departments based on content areas. This departmental structure provides structure and a shared discipline, but it also fosters the isolated nature of the disciplines, limiting the interdisciplinary opportunities students experience as they prepare for a twenty-first century workforce that has been transformed from the factory model to innovative multidisciplinary models.

THE GLOBALED 2 PROJECT

GlobalEd2 (GE2) is a set of interdisciplinary, problem-based simulations, and curricular supports intended to provide a venue for students to apply their developing knowledge and skills in an authentic, real-world activity. It is designed for ground teaching and learning in meaningful socioscientific contexts related to the world in which students currently live, representing an innovative approach to improving outcomes, particularly for high need, underrepresented students. Its targeted learning outcomes include increased engagement and knowledge across several disciplines, heighten positive affect around these domains, the development of STEM literacy, and improved college and career readiness skills (e.g., collaboration, problem solving, and written communication). Moreover, it is an evidenced-based curricular experience that has shown promise across multiple academic levels and a diverse array of students.[4]

GE2 evolved from the earlier model, GlobalEd, which was situated in the social sciences.[5] The current version has been developed through funding provided by the Institute of Educational Science in the US Department of Education to become an interdisciplinary learning environment centered on STEM literacy.[6] GlobalEd and GlobalEd 2 have serviced over

8000 middle grade through college students, and research studies across multiple implementations has demonstrated this approach to learning has high impact on a variety of important student outcomes, including writing argumentation and quality, science knowledge, interest in science, writing self-efficacy on STEM topics, problem solving, leadership, negotiation, academic motivation, and taking social perspective. Results have further demonstrated that these gains occur across diverse student groups, including Black, Latino, and female students. Finally, observations of implementations indicate changes in instructors' pedagogy consistent with problem-based learning (PBL). While GE2 has predominantly been implemented in middle school classrooms, it has also been successfully implemented in both high school and college level courses with similarly positive student outcomes.

GE2 is a technology-mediated curricular intervention, provided via a suite of web-based applications, including professional development (PD) and implementation support for instructors, resources, learning scaffolds for students, and a communications platform to enable collaborative interactions among students. The underlying technology provides consistency across implementations and scalability of the program to large number of students across multiple settings. A single simulation of GE2 can accommodate up to 20 classes of students ($n \sim 400$–500), and may be provided for a single institution or collaboratively across multiple institutions. Moreover, the technology infrastructure can handle multiple simultaneous simulations, affording delivery to an exponential number of students.

Previously, we have presented research data demonstrating GE2's specific impact on student writing, one of our strongest and most consistent outcomes across student settings (middle grade through college) and scenario topics (Water Resources and Climate Change).[7] In this chapter, we focus on our discussion on the impact of GE2 on students' self-efficacy, a belief in the self that is key to achieving educational goals. According to Albert Bandura and his research, a person's self-efficacy is very specific and tied to specific tasks and/or knowledge. It influences behavior by determining what the person attempts to achieve and the amount of effort applied to his/her performance. The psychological research literature of Bandura et al. has firmly established that self-efficacy is an important variable in predicting student engagement, motivation, task commitment, and learning outcomes. This research has demonstrated that self-efficacy is affected by cognitive, emotional, and behavioral variables which are tied to encouragement, challenge, previous success, and emotional arousal. If students' STEM experiences are successful, then their self-efficacy, and, in return their attitudes related to STEM, are augmented, increasing the likelihood of future engagement in the discipline. In contrast, when students' have unsuccessful STEM experiences, their associated attitudes are decreased, as they develop low self-efficacy for related content and tasks, and they are less likely to engage in STEM topics.

The following discussion will focus on how students' experiences in the GE2 simulations has positively affected their STEM self-efficacy.

STEM LITERACY

The science and academic communities have been sounding the warning alarm about the crisis in science education for years: Our schools are just not producing the STEM professionals necessary for the USA to maintain its scientific and technological prominence, thereby putting our current and future global economic standing at risk. Beyond the need for more highly trained professionals within STEM fields, however, we also face a much larger secondary societal crisis: The need to establish a scientifically literate citizenry that can make informed decisions at the local, regional, and national levels. Recent standardized test results indicate that only 21% of twelfth-graders performed at or above the proficient level in science, and our ranking internationally on the scientific literacy of our students, measured on tests like PISA, has rapidly fallen.[8] In order to engage with the many social, cultural, political, and ethical issues that arise from advances in knowledge, our population, not just our STEM professionals, needs to be sufficiently informed and efficacious with the principles of STEM. Issues related to global climate change, sources of alternative energy, evolution, and environmental preservation all require careful and informed decision-making by both citizens and elected leaders. Moreover, STEM literacy involves much more than just content knowledge; it also require an understanding of the representation and interpretation of scientific data, scientific explanations and projections, and the process of science. Further, STEM literacy involves cognitive and metacognitive abilities, collaborative teamwork, effective use of technology, and the abilities to engage in scientific discourse around global issues, synthesize disparate concepts, and persuade others to take informed action based on scientific evidence. These skills, the National Science Board has argued, may be even more important than knowing particular scientific facts.

Twenty-First Century Workforce Skills

STEM literacy skills parallel those employed in the authentic, socio-scientific work of twenty-first century scientists. Contemporary scientists need to be able to bring their knowledge, insights, and analytical skills to bear on matters of public importance. Often they can help the public and its representatives understand the likely causes of events (such as the potential for natural and technological disasters) and to estimate the potential effects of projected policies (such as the ecological impacts of various water conservation methods, as we are currently seeing in parts of the American West). In playing this advisory role, scientists are expected to be especially careful in distinguishing fact from interpretation and research findings from speculation and opinion in order to develop valid arguments, as are the citizens who are consuming this information to develop their own positions—the essence of a scientifically literate citizen.[9]

As such, argumentation is a central process necessary for the development of a scientifically literate citizenry. Argumentation includes any dialog that addresses "the coordination of evidence and theory to support or refute an explanatory conclusion, model, or prediction."[10] Research has demonstrated that when students engage in scientific argumentation, they not only learn to develop valid arguments but also learn science content while doing so. Further, there is convergent evidence that demonstrates that both instruction and authentic opportunities to write have been shown to improve writing skill.[11] While there has been strong advocacy for argumentation and writing in science, opportunities for students to learn how to engage in productive scientific argumentation in the current context of school-based science have been rare. This has been a driving force behind the emergence of college general education requirements of discipline-based writing experiences.

Through the work of O'Brien et al. research has also established a link between interest in science and science self-efficacy beliefs. It stands to reason, then, that if we can develop settings where students have the opportunity to experience success and illustrate the personal relevance of STEM topics in the world in which students live, we can positively impact their STEM self-efficacy and interest. As a result, we may better be able to affect student engagement and enrollment in the sciences with the outcomes of further developing their STEM literacies, thereby increasing the pool of viable candidates in the STEM workforce.

Many have argued that using the social sciences (i.e., psychology, anthropology, political science, economics, education, sociology) as a forum for integrating and applying science has the potential to develop a scientifically literate citizenry capable of bringing a scientific approach to bear on the practical, social, economic, and political issues of modern life. Furthermore, researchers, such as John Bransford et al. and David Jonassen, have illustrated for decades that leveraging interdisciplinary contexts, like the social sciences, provide opportunities for students to engage in real-world problem solving that can deepen students' understanding, their flexibility in the application of knowledge, and the transfer of knowledge to novel situations, while also reducing the likelihood of inert knowledge.

Socio-scientific contexts afford students the opportunity to ground their STEM learning in the world in which students currently live, making science personally relevant. Socio-scientific issues are complex and often do not have a single, clear-cut solution. Such issues confront students with situations in which they have to engage in formulating stances based on data, their own experiences and values, and collaborative decision-making. They are regarded as real-world problems that afford the opportunity for students to participate in the negotiation and development of meaning through scientific argumentation and promoting epistemic, cognitive, and social goals, as well as enhancing students' understanding of science.

To sustain our competitive edge in today's global economy, we must provide accessible and supportive pathways for *all* students to enroll in postsecondary education and complete their degrees in a timely fashion. Postsecondary education is the primary conduit for strengthening our workforce and ensuring a better quality of life for our citizens. Better educated people clearly have a greater chance than those who are less educated of obtaining secure jobs that provide opportunities for advancement, higher wages, greater health and retirement benefits, and greater opportunities in general.

Across several independent surveys of businesses and potential employers, the most commonly cited skills that industry requires in newly graduated college students include the abilities to solve complex, multidisciplinary problems, work successfully in teams, exhibit effective oral and written communication skills, and practice good interpersonal skills. However, industry leaders point out that many students who obtain their postsecondary degrees do not possess these skills and, as such, are not fully prepared to successfully participate in the twenty-first century workforce.[12] It seems that our education systems need to change—quickly!

THE GE2 PROJECT DESIGN PRINCIPLES

Current instructional practice, predominantly based within the cognitive perspective of learning, is at odds with research findings about how people learn with understanding. As stated in the 2012 report from the National Research Council, "Typical classroom activities convey either a passive and narrow view of learning or an activity oriented approach devoid of question-probing and only loosely related to conceptual learning goals."[13] Such instructional practices limit the teaching of high order thinking skills that are critical components of college and career readiness. Moreover, the transfer of learning resulting from course activities enacted in this way is also hindered, as there is little understanding of the contexts in which the acquired knowledge and skills are useful.

In light of the shortcomings of cognitive-based approaches to teaching, our theory of change is rooted within the sociocultural perspective on learning. The sociocultural perspective emerged in response to the perception that research and theory within the cognitive perspective was too narrowly focused on individual thinking and learning. In the sociocultural model, learning takes place as individuals participate in the practices of a community, using the tools, language, and other cultural artifacts of the community. From this perspective, learning is "situated" within and emerges from the practices in different settings and communities.

Problem-Based Learning

Problem-based learning is an enactment of sociocultural theory aimed at addressing the need for deep learning, the transfer of skills and knowledge, and situating learning. In contrast to more traditional teaching methods that use problems after theory has been introduced, PBL uses a problem scenario to initiate, focus, and motivate the learning of new concepts. PBL research has illustrated that knowledge needs to be *conditionalized*, that is, people should understand when and why knowledge is useful.[14] Further, the empirical evidence base examining PBL has illustrated that learning should be *contextualized*, or the learning environment should mirror the context in which the outcomes are expected to be utilized. Such conditionalization and contextualization demand that students interact with authentic, ill-structured problems—those where there is no one correct way to solve the problem and which require knowledge and skills from multiple topic areas or disciplines. PBL also includes a collaborative component; students often work in groups where collective decisions are made about task distribution, and in which group members investigate different aspects of the problem that together contribute to the total solution.

There is an extensive literature base examining the positive impact of PBL as a pedagogical approach for teaching across a large variety of domains and with a highly diverse array of students. Gains on important learning outcomes, including knowledge, affect, and the use of high order thinking skills, have been well chronicled. However, less well documented is the impact of PBL on more distant learning outcomes, such as academic progress and retention in college students. We identified only one study meeting the What Works Clearinghouse Evidence Standards, with reservations, examining this. Sabine Severiens and Henk Schmidt conducted a quasi-experimental study with 305 first-year Psychology students, examining academic progress/retention in terms of credit accrual. Comparing a fully implemented PBL approach to a conventional lecture-based approach and a mixed approach that integrated various forms of "active learning," results indicated that students who experienced the PBL pedagogy outpaced students in the other conditions with respect to persistence and the rate of credit accrual. Further, levels of social and academic integration were also higher among students in the PBL curriculum. While the research of Severiens and Schmidt provides initial evidence showing the promise of PBL to promote college success, larger scale work must be conducted to further explore PBL's full potential.

GlobalEd 2 and Problem-Based Learning Principles

GE2 is designed to meet the criteria outlined by Nick Zepkey and Linda Leach, as well as the high impact practices (HIPs) espoused by the AACU and George Kuh as requisite for engaging students at the postsecondary level. Moreover, GE2 is grounded in PBL principles and design components. These principles and their alignment in the GE2 design are presented in Table 1.

Description of the GlobalEd 2 Implementation

As described previously by Lawless et al. GE2 is a set of problem-based, online curricular activities that engages classes of students across multiple locations in simulated, multinational negotiations around a socio-scientific issue currently facing the world.[15] Within a single implementation of GE2,

Table 1   Principles of PBL related to the GE2 design

PBL principles	GlobalEd 2 implementation of the PBL principles
Anchoring learning to Problem Scenario	Problem Scenario provided in a global and multidisciplinary setting; Includes 4 issue areas in each team
Support learners in developing ownership and control over problem	Web-based application enables customization and learner-directed interactions; online informational resources provided; SimCon interactions to guide and prompt learners
Be based on ill-structured authentic, problems	Problem scenario based in real-world, global socio-scientific issues, e.g., water resources, climate change, food security
Be collaborative	Learners are required to collaborate within and across country teams with the goal of negotiating a multi-team agreement to address the problem
Provide alternative views and solutions	Social-perspective taking supported by issue areas, social, and cultural perspectives; international focus with SimCon monitoring and support
Require the students to reflect on both the content and the process	Debriefing phase is designed to promote reflection on the experience and to facilitate near and far transfer

16 to 18 classes participate, each assigned to represent the interests of a different country for the entire simulation. Students within each country are further broken down into four collaborative groups, called issue areas (e.g., Economics, Human Rights, Environment, and Health). These issue areas are consistent across all the classes in a simulation, enabling the students from one issue area to communicate with their counterparts in another class. Although negotiations may take place between the specific issue groups across countries, it is necessary that these four issue groups also negotiate within their class/country to reach a consensus to represent a unified policy stance.

At the beginning of a GE2 implementation, each participating class is presented with a problem scenario and the collective goal to reach an agreement with at least one other country. The scenario provides background information about a current issue in the world that requires the participating countries in the simulation to take timely action. It sets the common context for the countries in the simulation, anchoring interactions among students. Sample scenario topics include water resources, climate change, and food security. In addition, GE2 participants (students

& instructors) are supported by a set of three separate web applications: (1) the Student Research and Tools Database; (2) the Communications Platform, which hosts the online communications among students; and (3) the Instructor Portal for instructional support, scaffolding, and PD.

There are three phases of GE2. The first phase, the Research Phase, requires the students to use the online *Student Research and Tools Database* to learn about the issues presented in the problem scenario. Students must identify the key scientific issues of concern, as well as how their assigned country's culture, political system, geography, and economy influence their perspectives. Additionally, students also become familiar with the policies of the other countries included in the simulation in order to develop initial arguments and plan for potential collaborations. For example, in the water resources scenario, students use the *Student Research and Tools Database* to learn about water consumption, pollution, irrigation, and access to fresh, clean water, as well as other related issues currently facing each of the countries involved. Per the outcome of the Research Phase, students in each classroom work collaboratively to develop opening policy statements (written scientific arguments), containing their national position for each of the four issue areas and how they wish to start addressing the international problem presented in the scenario with other countries that they will also be negotiating within the simulation. These opening statements generally range in length from 400 to 900 words, though some detailed statements are longer. Opening statements are then shared as documents through the online *Communication Platform* and serve to launch Phase 2, the interactive negotiations among countries (student-to-student communications across teams).

Throughout the Interactive Phase, students work within their class to refine their arguments and negotiate international agreements with the other "countries," sharpening their arguments through the use of the *Student Research and Tools Database* and sharing them through the *Communication Platform*, in an asynchronous format similar to email. Based on prior implementations, the number of communications exchanged during the Interactive Phase can exceed 5000 (although length varies from a single sentence to multiparagraph exchanges). Students are also afforded the ability to engage in moderated synchronous conferences (i.e., like instant messaging) at various scheduled points throughout the Interactive Phase. These synchronous conferences are important for students to clarify understandings and push negotiations forward more quickly than is attainable through asynchronous communications.

In order to provide control and flow during the Interactive Phase, a trained simulation coordinator, "SimCon," monitors all e-messages among teams and facilitates the synchronous conferences. SimCon's role is similar to that of a virtual teacher/facilitator in an active learning class, in which SimCon oversees all aspects of the learning process and coaches students to think critically about the complex issues central to their written arguments. Further, SimCon monitors and provides feedback to students regarding the content (scientific and political), writing quality, and tone of their communications as a means of formative evaluation. SimCon's ability to moderate the dialogue and interactions among participating students is facilitated through a back-end control function in the *Communication Platform*.

The culminating event of the Interactive Phase is each country's closing statement, reflecting the final position of each country-team on the four issue areas. Students work collaboratively within their country-team issue area to construct these closing arguments, articulating points of agreement and topics where continued work is necessary among the participating countries. The posting of the closing statements in the *Communication Platform* marks the start of the third phase of the GE2 experience, Debriefing.

The Debriefing Phase is designed to activate metacognitive processes as students review what they learned and how they can apply this new knowledge and associated skills in other contexts and domains. SimCon facilitates a scheduled online debriefing conference through the *Communication Platform* with all participants, exploring issues related to learning outcomes, simulation processes, transfer, and feedback. Instructors are also trained to perform multiple debriefing activities to promote metacognition, learning, and transfer (e.g., examining local water issues or other tasks to relate the experience to the real world of environmental sustainability).

All interactions in GE2 are text-based—a purposive design for two reasons. First, the written artifacts students produce (e.g., opening/closing statements and negotiations) are a means of making students' thinking visible, providing an avenue for instructors and researchers to formatively assess students' engagement, scientific thinking, writing, leadership, and problem solving. Second, the use of this anonymous written communication mode allows educators to hold some factors in the educational context neutral (e.g., personal appearance, gender, race, and verbal accents). Students only identify themselves within GE2 as country, issue area, and their initials, for example, "ChinaEnvSWB," concealing their actual identities to students outside their specific class. As a result, typical stereotypes, associated with gender, race, or socioeconomic class, are minimized as factors influencing the interactions among participants.

Although GE2 is a technology-mediated experience, participation in the simulations only requires a device that is Internet capable, including netbooks, iPads/tablets, and smartphones. The platform-independent nature of GE2 provides access to the simulation almost anywhere, any time.

The role of instructors changes dramatically within GE2. Rather than being the traditional "knowledge bank" that simply transfers what they know to students, within GE2, instructors take on the role of learning guide. The instructor's role is not to inform the students but to encourage and facilitate opportunities for them to learn for themselves by using the provided problem scenario, simulation experience, and student-learning scaffolds as a focus for the learning. Instructors implementing GE2 are supported by both front-end and on-going PD provided through an online *Instructor Portal*. Prior to their first time in the role of GE2, instructors will take approximately 24 hours of online course in which they learn about GE2, the theory behind it, how teaching and assessment occurs within it, how to support students to write effectively, and the science and social science content needed to successfully implement it with students. In addition, weekly podcasts will be provided using a "just in time" training model, providing content and process to suggestions to instructors as demanded by the trajectory of the students' interactions in the simulation. Finally, an online learning community of instructors and GE2 staff is used as a forum for instructors across GE2 sites to exchange information, ask questions of each other and GE2 staff, and collaboratively develop new knowledge and resources about teaching with GE2.

The *Instructor Portal* also provides access to an array of GE2 web-based lesson plans and learning supports. The lessons are aimed at helping students to identify and align important information across disciplines that are relevant to the problem. Understanding the world water crisis, for example, requires that students understand the Earth's water purification cycle (hydrologic cycle), the economic implications of water trade, water as a "virtual" commodity, access to water as a human right, health issues, and water reclamation technologies. In addition to content, instructional materials are provided to help shape the quality of students' writing using a research-based approach.[16] Finally, examples of completed assignments and evaluation rubrics are provided to support assessment of student learning both formatively and summatively.

GlobalEd 2 in College Courses

GE2 is not a core curriculum in and of itself. Rather, it is a set of extended curricular activities that provides a venue for students to build and apply their knowledge and skills in an authentic problem space *in concert with* standard curricular practice. It is intended to deepen and strengthen, not replace, the understanding and use of the knowledge and skills that students develop from middle grade classes through college.

In college, GE2 aligns best with First-Year Experience (FYE) classes and has been taught by FYE instructors across multiple disciplines (i.e., from engineering to business and public health) in both the USA and abroad. As outlined in their book, *Striving for Excellence*, John Szarlan et al., outline the typical FYE learning objectives, including information literacy, academic writing, study skills, campus knowledge, understanding academic expectations, collaboration work, service learning, and problem solving. By engaging students with the content, their peers, and their instructor in an early college experience, GE2 allows students to take ownership of their learning and use of learning skills at the beginning of their postsecondary trajectory with the goal that they will apply these skills in other courses and experiences.

In the spring of 2012, we conducted a study of GE2 implementation in First-Year Experience (FYE) courses at a large northeastern public university. A total of 252 FYE students and their FYE instructors participated in a GE2 simulation on international water resources for an entire semester. The FYE course was a 1-credit course and met weekly for 60 minutes in class sections of 19 or fewer students throughout the 14-week semester (weeks 1 and 14 were reserved for assessments). Instructors of the course completed a training seminar to prepare them for implementing the curriculum with their students.

This study received IRB approval, and therefore participants were given the choice of whether to consent and be included in the research component of the class, which involved pre- and post-assessments. All students participated in the educational component of GE2. Consenting students completed a battery of pre-test prior to being introduced to GE2. Within this battery were two self-efficacy subscales and a social perspective taking scale. The Cronbach alpha reliability estimates for each of the three scales have exceeded 0.80 on previous samples[17] and were similar on the current sample. Once the students completed the pre-test battery, they began participation in the GE2 simulation, after which they completed the same assessments as post-assessments.

Following the pre-testing, students were informed of their assigned country, the scenario, and the four issue groups (e.g., Human Rights, Economics, Environment, and Health), which instructors allowed the students to select with the goal of creating roughly equivalent group sizes and gender distribution. There were 12 countries in the simulation, plus the USA, which was played by two GE2 staff members (which was not known to the FYE students). A veteran SimCon experienced in water resources and international affairs monitored all the online communications and hosted the synchronous conferences.

The data extracted during this study was examined to assess the impact of GE2 on the STEM self-efficacy of these college students and provide feedback on current features of its college implementations. Three specific research questions were addressed: whether there were gains from GE2 on students' (1) self-efficacy for educational technology, (2) general academic skills self-efficacy, and (3) the social perspective-taking skills.

A total of 252 college students (54% White, 28% Black, and 18% "other" or missing) participated in GE2 during the spring 2012 semester; 173 providing informed consent, with 101 providing matched pre- and post-data on our battery of assessments. A series of three separate paired t-tests were conducted on the pre- and post-measures of *Technology Use Self-efficacy*, *General Academic Self-efficacy*, and *Social Perspective Taking*. The results displayed in Table 2 demonstrate statistically significant increases from pre to post on all three measures. The results speak

Variable	T-statistic	Significance (p-value)
Technology Self-Efficacy	–2.365	0.023
Social Perspective Taking	–5.252	0.001
Academic Skills Self-Efficacy	–2.192	0.035

Table 2 FYE GE2 paired sample *t*-test results for pre- and post-testing

to the potential of PBL, and specifically GE2, as a meaningful context within which college students can experience twenty-first century skills and STEM content, as well as developing skills positively affecting their skills and STEM self-efficacy. Specifically, each of the three self-efficacy skills (Technology, Social Perspective Taking, and Academic Skills) were found to increase significantly in a simulated game of international negotiations on a STEM topic, water resources.

GE2 FOR INSTRUCTORS AND STUDENTS

Over five years of research on GE2 have focused on two groups of end users: Instructors and their students. For instructors, GE2 promotes a shift of their pedagogical practices away from a traditional approach of being a content expert in a particular domain who controls the flow of the class, lectures, and/or transmits information.[18] In GE2, instructors are *guides* who facilitate a *student-centered* learning approach. Instructors are not content experts across the multiple domains represented, but serve as *model knowledgeable information seekers and evaluators*. GE2 not only creates a new innovative approach to teaching with PBL, but also trains and supports teachers on the enactment of PBL in their classes prior to, and through, the entirety of the simulation.

With respect to students, GE2 engages learners, helping them to develop their STEM self-efficacy and STEM literacy (knowledge, skills, and attitudes that every citizen needs to know), as well as college and career readiness skills. GE2 also places a pronounced emphasis on the development of students' written communication, discussed in other forums, integrating a research-based instructional framework for writing to foster the development of written communication skills.[19] Beyond just learning written communication, there is also substantial evidence indicating that writing is also an effective tool for enhancing knowledge acquisition and cognitive skill development in the disciplines, student affect, and engagement.

The nature of the GE2 simulation also requires that teams work together, representing countries across issue areas and collaborating with other country teams across the large simulation space. Engaging students in these collaborative activities is the mechanism through which *team building* and *cooperation skills* are developed. Through the give and take of negotiations within the simulation, students engage in developing *problem solving skills* as they learn the complexity of the problem space, separate relevant from irrelevant information, and apply various tactics and heuristics to gain traction and progress toward their goal of agreement with at least one other country.

While each of the above student outcomes is important individually, in aggregate, GE2 fosters the much broader outcome of students' engagement with other students, their instructor, and the content, as well as intellectual development.

CONCLUSION

GE2 is grounded in empirical research findings drawn from multiple fields that influence STEM education, including the following:

- If students' STEM experiences are unsuccessful, then their STEM self-efficacy is diminished, decreasing the likelihood of future engagement in the discipline.
- The choice to enroll in STEM courses and pursue STEM-related occupations is mediated by a student's STEM-based self-efficacy. Low self-efficacy yields low engagement.
- Leveraging interdisciplinary contexts, like the social sciences, as a venue to engage in real-world problem solving can deepen students' understanding, flexibility in application, and transfer of knowledge.
- Embedding STEM curricula in global socio-scientific issues is a means for opening up science to females and excluded or disadvantaged ethnic and class groups.
- Scientific argumentation is a central STEM literacy. When students engage in scientific argumentation, they not only learn to develop valid arguments but also learn science content while they do so.
- Writing instruction and practice writing for authentic audiences improve writing skill.

A better understanding of how to maintain and cultivate middle school through college students' interest in STEM education and careers paths is vital to addressing the STEM pipeline issues and STEM literacy in the USA. The instructional approach proposed by GE2 not only addresses this need, but also broadens the focus on what, where, and how STEM literacy can be cultivated, enhanced, and assessed. Nearly 15 years of research and development, from the first iteration of GlobalEd to the current version of the STEM based GlobalEd 2 Project, has yielded consistently positive student (middle grades through college) learning results, including increased STEM self-efficacy, increased knowledge in both the social sciences and STEM fields, increased writing skills, and increased student engagement and motivation.

In transitioning GE2 from a successful research intervention to a viable educational curriculum designed to promote important student learning outcomes, we have determined that the human resources necessary to implement GE2 are modest. At scale, calculations indicate that GE2 can run at less than $25/student for veteran GE2 instructors and less than $40/student for novice/first time GE2 instructors (those requiring initial training) for the middle grades through college. This equates to a total of $500–$800 per class of 20 students—less than the average tuition postsecondary institutions charge for an individual student taking a three-credit course, even by conservative estimates, and less than the costs of classroom books in secondary schools. With our instructor training provided completely online for the last five years, it is very clear that GE2 can be brought to scale both effectively and efficiently for middle schools through colleges. Furthermore, the curriculum implementation may be adjusted to meet the needs of the educational environment, varying the implementations from 6 weeks to 14 weeks, while adapting the required amount of time per week for students, both in traditional settings, as well as virtually. Therefore, GE2 is both powerful and adaptable, adept at meeting the goals of educational institutions, their instructors, and their students.

While we are greatly encouraged by the results of studies of the GE2 approach, there remains much more to learn about its direct and long-term impact on student learning, as well as why PBL, and specifically GE2, enhances student knowledge, skills, and attitudes, so that we may advance student learning. Nevertheless, the evidence supports GE2 as an effective, cost-efficient approach to education that improves students' STEM competencies, resulting in more knowledgeable citizens who are ready to engage with the complexities and ramifications of science and the policies that shape it.

NOTES

1. Derek Bok, *Our Underachieving Colleges: A Candid Look at How Much Students Learn and Why They Should be Learning More* (Princeton, NJ: Princeton University Press, 2007), 226.

2. John Morely, "Labour Market Developments in the New EU Member States," *Industrial Relations Journal* 38, no. 6 (2007): 458–79.
3. John J. Heldrich, "Survey of New Jersey Employers to Assess the Ability of Higher Education Institutions to Prepare Students for Employment," *The New Jersey Commission on Higher Education*, 2005, accessed June 10, 2015, http://files.eric.ed.gov/fulltext/ED485290.pdf.
4. Kimberly A. Lawless and Scott W. Brown, "Developing Scientific Literacy Skills through Interdisciplinary, Technology-based Global Simulations: GlobalEd2," *The Curriculum Journal* 1 (2015): 1–22.
5. See Mark A. Boyer and Scott W. Brown, "Gender, Technology and Group Decision-Making: An Experimental Study in Secondary Education International Studies," U.S. Department of Education; Office of Educational Research and Improvement—Field Initiated Grants: ED-ERI-84.30ST, 2001; Mark A. Boyer, Scott W. Brown, Michael Butler, Natalie Florea, Maria Hernandez, Paula R. Johnson, Lin Meng, and Clarisse Lima, "Educating for Global Awareness: Implications for Governance and Generational Change," *Global Change, Peace & Security* 16, no. 1 (2004): 73–77; Scott W. Brown, Mark A. Boyer, Hayley Mayall, Paula R. Johnson, Lin Meng, Michael Butler, Katherine Weir, Natalie Florea, Maria Hernandez, and Sally Reis, "The GlobalEd Project: Gender Differences in a Problem-based Learning Environment of International Negotiations," *Instructional Science* 31, nos. 4–5 (2003): 255–76; Natalie Florea, Mark A. Boyer, Scott W. Brown, Michael Butler, Maria Hernandez, Katherine Weir, Lin Meng, Paula R. Johnson, Hayley Mayall, and Clarisse Lima, "Negotiating from Mars to Venus: Some Findings on Gender's Impact in Simulated International Negotiations," *Simulation and Games* 34, no. 2 (2003): 226–48.
6. The following federal grants supporting GlobalEd 2: Scott W. Brown, Kimberly A. Lawless, and Mark A. Boyer, "Expanding the Science and Literacy Curricular Space: The GlobalEd II Project," U.S. Department of Education: The Institute of Education Sciences, IES, #R305A080622, 2008; Scott W. Brown and Kimberly A. Lawless, "GlobalEd 2: Efficacy and Replication—Goal 3," U.S. Department of Education: The Institute of Education Sciences, IES, #R305A1300195, 2013.

7. For a review of the findings from GlobalEd 2 research, see Scott W. Brown and Kimberly A. Lawless, "Promoting Students' Writing Skills in Science through an Educational Simulation: The GlobalEd 2 Project," In *Human-Computer interaction, Part I, HCII 2014*, edited by Panayiotis Zaphiris, LNCS 8523, 371–79, Switzerland: Springer International Publishing, 2014; Scott W. Brown, Kimberly A. Lawless, and Mark A. Boyer, "The GlobalEd 2 Simulations: Promoting Positive Academic Dispositions in Middle School Students in a Web-based PBL Environment," in *Essential Readings in Problem-based Learning*, eds. Andrew Walker, Heather Leary, Cindy Hmelo-Silver and Peggy Ertmer (West Lafayette, IN: Purdue University Press, 2015), 147–59; Scott W. Brown, Kimberly A. Lawless, and Mark A. Boyer, "Promoting Science Literacy in College Students" (presented at the International Convention of Psychological Science, Amsterdam, NL), accessed March 14, 2015, https://osf.io/8zh59/; Scott W. Brown, Kimberly A. Lawless, Mark A. Boyer, Andrew Cutter, Kamila Brodowinska, Daniel O'Brien, Gregory Williams, Nicole Powell, and Maria Fernada Enriquez, "GlobalEd 2: Using PBL to Promote Learning in Science and Writing" (poster presented at the Association of Psychological Science Conference, Boston, MA, May 2010); Scott W. Brown, Kimberly A. Lawless, Mark A. Boyer, Gregory P. Mullin, Mariya Yukhymenko, Andrew Cutter, Kamila Brodowinska Bruscianelli, Nicole Powell, Maria Fernada Enriquez, Gerald Rice, and Gena K. Khodos, "Impacting Middle School Students' Science Knowledge with Problem-based Learning Simulations," (paper presented at the IADIS International Conference Cognition and Exploratory Learning in Digital Age, Rio de Janeiro, Brazil, November 2011); Hunter Gehlbach, Scott W. Brown, Andri Ioannou, Mark A. Boyer, Natalie F. Hudson, Anat Niv-Solomon, Donalyn Maneggia, and Laua Janik, "Increasing Interest in Social Studies: Social Perspective Taking and Self-efficacy in Stimulating Simulations" *Contemporary Educational Psychology*, 33, (2008): 894–914; Kimberly A. Lawless and Scott W. Brown, "Developing Scientific Literacy Skills through Interdisciplinary, Technology-based Global Simulations: GlobalEd2," *The Curriculum Journal* 1 (2015): 1–22.; Kimberly A. Lawless, Scott W. Brown, Kamila Brodowinska, Kathryn Field, Lisa Lynn, Jeremy Riel, Lindsey Le-Gervais, Charles Dye, and Rasis Alanazi, "Expanding the Science and Literacy Curricular Space: The

GlobalEd 2 Project" (paper presented at Eastern Educational Research Association Annual Meeting, Jacksonville, FL, February, 2014); Kimberly A. Lawless, Scott W. Brown, Kamila Brodowinska, Lisa Lynn, Jeremy Riel, Kathryn Fields, Lindsey Le-Gervais, and Gregory P. Mullin, "The GE2 Project—Developing a Scientifically Literate Citizenry," In *Encyclopedia of Information Science and Technology*, Third edition, ed. Mehdi Khosrow-Pour, Hershey, PA: IGI Global, 2014.

8. National Assessment of Educational Progress (NAEP) 2009, accessed April 22, 2012, http://www.nagb.org/publications/frameworks/science-09.pdf; National Center for Educational Statistics (NCES) 2011, accessed April 27, 2012, http://nces.ed.gov/pubs2011/2011004.pdf.

9. National Research Council, "Education for Life and Work: Developing Transferable Knowledge and Skills in the 21st Century" (Washington, DC: National Academy Press, 2012).

10. Jonathan F. Osborne, Sibel Erduran, and Shirely Simon, *Ideas, Evidence and Argument in Science. Inservice Training Pack, Resource Pack and Video* (London: Nuffield Foundation, 2004), 995.

11. See Jay Lemke, *Talking Science: Language, Learning, and Values* (Norwood, NJ: Ablex, 1990); Victor Sampson and Douglas Clark, "Assessment of the Ways Students Generate Arguments in Science Education: Current Perspectives and Recommendations for Future Directions," *Science Education*, 92, (2008): 447–472; Philip H. Scott, "Teacher Talk and Meaning Making in Science Classrooms: A Vygotskian Analysis and Review," *Studies in Science Education* 32 (1998): 45–80.

12. American Association of Colleges and Universities (AACU), "Raising the Bar: Employers' Views on Colleges Learning in the Wake of the Economic Downturn," 2010, accessed May 11, 2015, https://www.aacu.org/sites/default/files/files/LEAP/2009_EmployerSurvey.pdf; Gallup Postsecondary Education Aspirations and Barriers, 2014, accessed June 9, 2015, http://www.luminafoundation.org/resources/postsecondary-education-aspirations-and-barriers; John Morely, "Labour market developments in the new EU Member States," *Industrial Relations Journal* 38, no. 6 (2007): 458–79.

13. National Research Council, *Education for Life and Work: Developing Transferable Knowledge and Skills in the 21st Century* (Washington, DC: National Academy Press, 2012), 253.

14. Janet L. Kolodner, *Case-Based Reasoning* (San Mateo, CA: Morgan Kaufmann, 1993).

15. Lawless and Brown, "Developing Scientific Literacy Skills through Interdisciplinary, Technology-based Global Simulations."; Brown, Lawless, and Boyer, "The GlobalEd 2 Simulations."

16. Katherine L. McNeill and Joseph Krajcik, "Scientific Explanations: Characterizing and Evaluating the Effects of Teachers' Instructional Practices on Student Learning," *Journal of Research in Science Teaching* 45, no. 1 (2008): 53–78.

17. Scott W. Brown, Kimberly A. Lawless, and Mark A. Boyer, "The GlobalEd 2 Project: Expanding the Science and Literacy Curricular Space." In *Proceedings of World Conference on E-Learning in Corporate, Government, Healthcare, and Higher Education,* ed. Theo Bastiaens, Jon Dron and Cindy Xin, 160–4 (Chesapeake, Va.: AACE, 2009) http://www.editlib.org/p/32449.

18. Refer to the following two studies by Brodowinska et al describing the PBL fidelity factors: Kamila Brodowinska, Kimberly A. Lawless, Mark A. Boyer, Scott W. Brown, Daniel O'Brien, Gregory Williams, Nicole Powell, and Maria Fernada Enriquez, "Teachers' Approaches to Implementing a Problem-based Learning Simulation" (paper presented at the Society for Information Technology & Teacher Education Conference, San Diego, CA, April 2010); Kamila Brodowinska, Lisa Lynn, Kimberly A. Lawless, Scott W. Brown, Mark A. Boyer, Daniel O'Brien, Andrew Cutter, Maria Fernada Enriquez, Gena Khodos, Donalyn Maneggia, Gregory P. Mullin, Nicole Powell, and Gregory Williams, "Teachers' Varied Approaches to Implementing a PBL, GlobalEd 2 Simulation: An Evolved Analysis" (paper presented at the American Educational Research Conference, Vancouver, Canada, April 2012).

19. Lawless and Brown, "Developing Scientific Literacy Skills through Interdisciplinary, Technology-based Global Simulations."; Brown, Lawless, and Boyer, "The GlobalEd 2 Simulations."

Acknowledgements This research has been supported by grants from the US Department of Education' Office of Educational Research and Improvement (OERI), and the Institute for Education Sciences (IED). The authors gratefully acknowledge the support of OERI and IES for the research reported in this chapter. The opinions and positions expressed in this chapter represent those of the authors and do not necessarily represent the position of the US Department of Education.

BIBLIOGRAPHY

Alliance for Excellent Education. 2011. Saving now and saving later: How high school reform can reduce the nation's wasted remediation dollars. May 2011. Accessed April 16, 2014. http://all4ed.org/reports-factsheets/saving-now-and-saving-later-how-high-school-reform-can-reduce-the-nations-wasted-remediation-dollars/.

American Association of Colleges and Universities (AAC&U). 2010. Raising the bar: Employers' views on colleges learning in the wake of the economic downturn. Accessed May 11, 2015. https://www.aacu.org/sites/default/files/files/LEAP/2009_EmployerSurvey.pdf.

Anderson, Ronald D. 2002. Reforming science teaching: What research says about inquiry. *Journal of Science Teacher Education* 13: 1–12.

Bandura, Albert. 1986. *Social foundations of thought and action: A social cognitive theory*. Englewood Cliffs: Prentice-Hall.

Bandura, Albert. 1997. *Self-efficacy: The exercise of control*. Englewood Cliffs: Prentice-Hall.

Bangert-Drowns, Robert L., Marlene M. Hurley, and Barbara Wilkinson. 2004. The effects of school-based writing-to-learn interventions on academic achievement: A meta-analysis. *Review of Educational Research* 74: 29–58.

Bednar, Anne K., Donald J. Cunningham, Thomas M. Duffy, and J. David Perry. 1992. Theory into practice: How do we link? In *Constructivism and the technology of instruction: A conversation*, ed. Thomas M. Duffy and David J. Jonassen, 17–35. Hillsdale: Lawrence Erlbaum Associates.

Bereiter, Carl, and Marlene Scardamalia. 1987. *The psychology of written composition*. Hillsdale: Lawrence Erlbaum Associates.

Bok, Derek. 2007. *Our underachieving colleges: A candid look at how much students learn and why they should be learning more*. Princeton: Princeton University Press.

Bowman, Nicholas A. 2011. Promoting participation in a diverse democracy: A meta-analysis of college diversity experiences and civic engagement. *Review of Educational Research* 81(1): 29–68.

Boyer, Mark A., and Scott W. Brown. 2011. *Gender, technology and group decision-making: An experimental study in secondary education international studies*. US Department of Education; Office of Educational Research and Improvement–Field Initiated Grants: ED-ERI-84.30ST.

Boyer, Mark A., Scott W. Brown, Michael Butler, Natalie Florea, Maria Hernandez, Paula R. Johnson, Meng Lin, and Clarisse Lima. 2004. Educating for global awareness: Implications for governance and generational change. *Global Change, Peace & Security* 16(1): 73–77.

Bransford, John D., Nancy Vye, Lea T. Adams, and Greg A. Perfetto. 1989. Learning skills and the acquisition of knowledge. In *Foundations for a psychology of education*, ed. Alan Lesgold and Robert Glaser, 199–249. Hillsdale: Lawrence Erlbaum Associates.

Brodowinska, Kamila, Kimberly A. Lawless, Mark A. Boyer, Scott W. Brown, Daniel O'Brien, Gregory Williams, Nicole Powell, and Maria Fernada Enriquez. 2010. Teachers' approaches to implementing a problem-based learning simulation. Paper presented at the Society for Information Technology & Teacher Education conference, San Diego, CA, April.

Brodowinska, Kamila, Lisa Lynn, Kimberly A. Lawless, Scott W. Brown, Mark A. Boyer, Daniel O'Brien, Andrew Cutter, Maria Fernada Enriquez, Gena Khodos, Donalyn Maneggia, Gregory P. Mullin, Nicole Powell, and Gregory Williams. 2012. Teachers' varied approaches to implementing a PBL, GlobalEd 2 simulation: An evolved analysis. Paper presented at the American Educational Research Conference, Vancouver, Canada, April.

Brown, Scott W., and Frederick B. King. 2000. Constructivist pedagogy and how we learn: Educational psychology meets international studies. *International Studies Perspectives* 1: 245–253.

Brown, Scott W., and Kimberly A. Lawless. 2013. GlobalEd 2: Efficacy and replication—Goal 3. US Department of Education: The Institute of Education Sciences, IES. #R305A1300195.

Brown, Scott W., and Kimberly A. Lawless. 2014. Promoting students' writing skills in science through an educational simulation: The GlobalEd 2 project. In *Human-computer interaction, Part I, HCII 2014*, LNCS 8523, ed. Panayiotis Zaphiris, 371–379. Switzerland: Springer International Publishing.

Brown, Scott W., Mark A. Boyer, Hayley Mayall, Paula R. Johnson, Lin Meng, Michael Butler, Katherine Weir, Natalie Florea, Maria Hernandez, and Sally Reis. 2003. The GlobalEd project: Gender differences in a problem-based learning environment of international negotiations. *Instructional Science* 31(4–5): 255–276.

Brown, Scott W., Mark A. Boyer, Paula R. Johnson, Clarisse Lima, Michael Butler, Natalie Florea, and Jeremy Rich. 2004. The GlobalEd project: Problem-solving and decision making in a web-based PBL. *Proceedings of Ed-Media 2004, the World Conference on Educational Media and Technology*, 1967–1973. Ed-Media: Lugano, Switzerland.

Brown, Scott W., Kimberly A. Lawless, and Mark A. Boyer. 2005. Promoting science literacy in college students. Presented at the International Convention of

Psychological Science, Amsterdam, NL. Accessed March 14, 2015. https://osf. io/8zh59/.

Brown, Scott W., Kimberly A. Lawless, and Mark A. Boyer. 2008. *Expanding the science and literacy curricular space: The GlobalEd II project.* US Department of Education: The Institute of Education Sciences, IES. #R305A080622.

Brown, Scott W., Alexander Lyras, Andri Nicolaou, Mark A. Boyer, P. Polyviou, Eleni Kotziamani, Anat Niv-Solomon, Laura Janik, Hunter Gehlbach, and Donalyn Maneggia. 2008. Problem-solving, decision-making, and negotiations in an interdisciplinary environment: The 2007 Doves GlobalEd project. Paper presented at the American Educational Research Association Conference, New York City, NY, March 2008.

Brown, Scott W., Kimberly A. Lawless, and Mark A. Boyer. 2009. The GlobalEd 2 project: Expanding the science and literacy curricular space. In *Proceedings of world conference on e-learning in corporate, government, healthcare, and higher education*, ed. Theo Bastiaens, Jon Dron, and Cindy Xin, 160–164. Chesapeake: AACE. http://www.editlib.org/p/32449.

Brown, Scott W., Kimberly A. Lawless, Mark A. Boyer, Andrew Cutter, Kamila Brodowinska, Daniel O'Brien, Gregory Williams, Nicole Powell, and Maria Fernada Enriquez. 2010. GlobalEd 2: Using PBL to promote learning in science and writing. Poster presented at the Association of Psychological Science Conference, Boston, MA, May 2010.

Brown, Scott W., Kimberly A. Lawless, Mark A. Boyer, Gregory P. Mullin, Mariya Yukhymenko, Andrew Cutter, Kamila Brodowinska Bruscianelli, Nicole Powell, Maria Fernada Enriquez, Gerald Rice, and Gena K. Khodos. 2011. Impacting middle school students' science knowledge with problem-based learning simulations. Paper presented at the IADIS International Conference Cognition and Exploratory Learning in Digital Age, Rio de Janeiro, Brazil, November 2011.

Brown, Scott W., Kimberly A. Lawless, Mark A. Boyer, Mariya Yukhymenko, Gregory P. Mullin, Kamila Brodowinska, Gena Khodos, Nicole Powell, and Lisa Lynn. 2012. Increasing technology and writing self-efficacy through a PBL simulation: GlobalEd 2. Paper presented at the Eastern Educational Research Association Conference; Hilton Head, SC, February 2012.

Brown, Scott W., Kimberly A. Lawless, and Mark A. Boyer. 2015. The GlobalEd 2 simulations: Promoting positive academic dispositions in middle school students in a web-based PBL environment. In *Essential readings in problem-based learning*, ed. Andrew Walker, Heather Leary, Cindy Hmelo-Silver, and Peggy Ertmer, 147–159. West Lafayette: Purdue University Press.

Brown, Scott W., Kimberly A. Lawless, and the GlobalEd 2 Team. 2015. Promoting conceptual change in science through a PBL simulation: GlobalEd 2. Presented at the annual meeting of the Association of Psychological Science, New York City, NY, May 23, 2015.

Carlone, Heidi B. 2004. The cultural production of science in reform-based physics: Girls' access, participation, and resistance. *Journal of Research in Science Teaching* 41(4): 392–414.

Cavagnetto, Andy R. 2010. Argument to foster scientific literacy: A review of argument interventions in K-12 science contexts. *Review of Educational Research* 80(3): 336–371.

Chinn, Clark, and Betina Malhotra. 2002. Epistemologically authentic inquiry in schools: A theoretical framework for evaluating inquiry tasks. *Science Education* 86(2): 175–218.

DeHart Hurd, Paul. 1998. Science literacy: New minds for a changing world. *Science Education* 82: 407–418.

Duschl, Richard A., and Jonathan Osborne. 2002. Supporting and promoting argumentation discourse in science education. *Studies in Science Education* 38: 39–72.

Eisenhart, Margaret A., and Elizabeth Finkel. 1998. *Women's science: Learning and succeeding from the margins*. Chicago: University of Chicago Press.

Florea, Natalie, Mark A. Boyer, Scott W. Brown, Michael Butler, Maria Hernandez, Katherine Weir, Lin Meng, Paula R. Johnson, Hayley Mayall, and Clarisse Lima. 2003. Negotiating from Mars to Venus: Some findings on gender's impact in simulated international negotiations. *Simulation and Games* 34(2): 226–248.

Gallup. 2014. Postsecondary education aspirations and barriers. Accessed June 9, 2015. http://www.luminafoundation.org/resources/postsecondary-education-aspirations-and-barriers.

Gehlbach, Hunter, Scott W. Brown, Andri Ioannou, Mark A. Boyer, Natalie F. Hudson, Anat Niv-Solomon, Donalyn Maneggia, and Laua Janik. 2008. Increasing interest in social studies: Social perspective taking and self-efficacy in stimulating simulations. *Contemporary Educational Psychology* 33: 894–914.

Goodnough, Karen C., and Woei Hung. 2008. Engaging teachers' pedagogical content knowledge: Adopting a nine-step problem-based learning model. *Interdisciplinary Journal of Problem-Based Learning* 2(2): 61–90.

Graham, Steve, and Dolores Perin. 2007. Writing next: Effective strategies to improve writing of adolescents in middle and high schools. Alliance for Excellent Education. Accessed March 18, 2014. http://carnegie.org/fileadmin/Media/Publications/PDF/writingnext.pdf.

Greening, Toni. 1998. Scaffolding for success in problem-based learning. *Medical Education Online* 3(4): 1–15.

Harackiewicz, Judith, Kenneth Barron, Paul Pintrich, Andrew Elliot, and Todd Thrash. 2002. Revision of achievement goal theory: Necessary and illuminating. *Journal of Educational Psychology* 94(6): 38–645.

Hargreaves, Andy, and Moore Shawn. 2000. Curriculum integration and classroom relevance: A study of teachers' practice. *Journal of Curriculum and Supervision* 15(2): 89–112.

Hayesm, John R. 2000. A new framework for understanding cognition and affect in writing. In *Perspectives on writing: Research, theory and practice*, ed. Roselmina Indrisano and James R. Squire, 6–44. Newark: International Reading Association.

Heldrich, John J. 2005. Survey of New Jersey employers to assess the ability of higher education institutions to prepare students for employment. The New Jersey Commission on Higher Education. Accessed June 10, 2015. http://files.eric.ed.gov/fulltext/ED485290.pdf.

Hmelo-Silver, Cindy E. 2004. Problem-based learning: What and how do students learn? *Educational Psychology Review* 16(3): 235–266.

Holliday, William, Larry D. Yore, and Donna E. Alvermann. 1994. The reading-science-learning-writing connection: Breakthroughs, barriers, and promises. *Journal of Research in Science Teaching* 31(9): 877–893.

Jiménez-Aleixandre, Maria-Pilar. 2002. Knowledge producers or knowledge consumers? Argumentation and decision-making about environmental management. *International Journal of Science Education* 24: 1171–1190.

Johnson, Paula R., Mark A. Boyer, and Scott W. Brown. 2011. Vital interests: Cultivating global competence in the international studies classroom. *Globalisation, Societies and Education* 9(3–4): 503–519.

Jonassen, David H. 2010. Assembling and analyzing the building blocks of problem-based learning environments. In *Handbook of improving performance in the workplace, volume one: Instructional design and training delivery*, ed. Kenneth H. Silber and Wellesley R. Foshay, 361–394. San Francisco: Pfeiffer.

Kolodner, Janet L. 1993. *Case-based reasoning*. San Mateo: Morgan Kaufmann.

Koschmann, Timothy, Ann C. Kelson, Paul J. Feltovich, and Howard S. Barrows. 1996. Computer-supported problem-based learning: A principled approach to the use of computers in collaborative learning. In *CSCL: Theory and practice of an emerging paradigm*, ed. Timothy Koschmann, 83–124. Mahwah: Lawrence Erlbaum.

Kuh, George D. 2008. *High-impact educational practices: What they are, who has access to them, and why they matter*. Washington, DC: AAC&U.

Lave, Jean, and Etienne Wenger. 1991. *Situated learning: Legitimate peripheral participation*. Cambridge: Cambridge University Press.

Lawless, Kimberly A., and Scott W. Brown. 2015. Developing scientific literacy skills through interdisciplinary, technology-based global simulations: GlobalEd2. *The Curriculum Journal* 1: 1–22. doi:10.1080/09585176.2015.1009133.

Lawless, Kimberly A., Scott W. Brown, Mark A. Boyer, Kamila Brodowinska, Gregory P. Mullin, Mariya Yukhymenko, Gena Khodos, Lisa Lynn, Andrew

Cutter, Nicole Powell, and Maria Fernada Enriquez. 2011. Expanding the science and writing curricular space: The GlobalEd2 project. Paper presented at the IADIS International Conference Cognition and Exploratory Learning in Digital Age, Rio de Janeiro, Brazil, November, 2011.

Lawless, Kimberly A., Scott W. Brown, Mark A. Boyer, Kamila Brodowinska, Lisa Lynn, Gregory P. Mullin, and Mariya Yukhymenko. 2012. Developing scientific literacy skills through interdisciplinary, technology-based global simulations: GlobalEd 2. Paper presented at the Eastern Educational Research Association Conference, Hilton Head, SC, February 2012.

Lawless, Kimberly A., Scott W. Brown, Kamila Brodowinska, Kathryn Field, Lisa Lynn, Jeremy Riel, Lindsey Le-Gervais, Charles Dye, and Rasis Alanazi. 2014. Expanding the science and literacy curricular space: The GlobalEd 2 project. Paper presented at Eastern Educational Research Association annual meeting, Jacksonville, FL, February, 2014.

Lawless, Kimberly A., Scott W. Brown, Kamila Brodowinska, Lisa Lynn, Jeremy Riel, Kathryn Fields, Lindsey Le-Gervais, and Gregory P. Mullin. 2014. The GE2 project—Developing a scientifically literate citizenry. In *Encyclopedia of information science and technology*, 3rd ed, ed. Mehdi Khosrow-Pour. Hershey: IGI Global.

Leary, Heather, Andrew Walker, Brett E. Shelton, and Harrison Fitt. 2015. Exploring the relationships between tutor background, tutor training, and student learning: A problem-based learning meta-analysis. In *Essential readings in problem-based learning*, ed. Andrew Walker, Heather Leary, Cindy Hmelo-Silver, and Peggy Ertmer, 331–354. West Lafayette: Purdue University Press.

Lemke, Jay. 1990. *Talking science: Language, learning, and values.* Norwood: Ablex.

Levinson, Ralph and Sheila Turner, eds. 2001. *The teaching of social and ethical issues in the school curriculum, arising from developments in biomedical research: A research study of teachers.* London: Institute of Education. Accessed February 15, 2013. http://www.wellcome.ac.uk/stellent/groups/corporatesite/@msh_peda/documents/web_document/wtd003444.pdf.

McNeill, Katherine L., and Joseph Krajcik. 2008. Scientific explanations: Characterizing and evaluating the effects of teachers' instructional practices on student learning. *Journal of Research in Science Teaching* 45(1): 53–78.

Millar, Robin, and Jonathan F. Osborne (eds.). 1998. *Beyond 2000: Science education for the future.* London: King's College, School of Education.

Monk, Martin, and Jonathan F. Osborne. 1997. Placing the history and philosophy of science on the curriculum: A model for the development of pedagogy. *Science Education* 81: 405–424.

Morely, John. 2007. Labour market developments in the new EU member states. *Industrial Relations Journal* 38(6): 458–479.

National Assessment of Educational Progress (NAEP). 2009. http://www.nagb.org/publications/frameworks/science-09.pdf. Accessed 22 Apr 2012.

National Center for Educational Statistics (NCES). 2011. Accessed April 27, 2012. http://nces.ed.gov/pubs2011/2011004.pdf.

National Research Council. 1996. *National science education standards.* Washington, DC: National Academy Press.

National Research Council. 2012. *Education for life and work: Developing transferable knowledge and skills in the 21st century.* Washington, DC: National Academy Press.

National Science Board. 2008. Research and development: Essential foundations for U.S. competitiveness in a global economy. Accessed April 4, 2012. http://www.nsf.gov/statistics/nsb0803/nsb0803.pdf.

Newcombe, Nora S., Nalini Ambady, Jacquelynne Eccles, Louis Gomez, David Klahr, Marcia Linn, Kevin Miller, and Kelly Mix. 2009. Psychology's role in mathematics and science education. *American Psychologist* 64(6): 538–550.

Niv-Solomon, Anat, Laura Janik, Mark A. Boyer, Natalie F. Hudson, Brian Urlacher, Scott W. Brown, and Donalyn Maneggia. 2011. Evolving beyond self-interest? Some experimental findings from simulated international negotiations. *Simulation and Gaming* 42(6): 711–732.

O'Brien, Virginia, Manuel Martinez-pons, and Mary Kopala. 1999. Mathematics self-efficacy, ethnic identity, gender, and career interests related to mathematics and science. *Journal of Educational Research* 92(4): 231–235.

Osborne, Jonathan F., Sibel Erduran, Shirely Simon, and Martin Monk. 2001. Enhancing the quality of argument in school science. *School Science Review* 82(301): 63–70.

Osborne, Jonathan F., Sibel Erduran, and Shirely Simon. 2004. *Ideas, evidence and argument in science. Inservice training pack, resource pack and video.* London: Nuffield Foundation.

Partnership for 21st Century Skills. Accessed June 1, 2015. http://www.nea.org/home/34888.htm.

Rivard, Leonard P., and Stanley B. Straw. 1994. The effect of talk and writing on learning science: An exploratory study. *Science Education* 84(5): 566–593.

Robinson, Shane J., Bryan Garton, and Paul R. Vaughn. 2007. Becoming employable: A look at graduates' and supervisors' perceptions of the skills needed for employability. *NACTA Journal* 51(2): 19–26.

Rogoff, Barbara. 1990. *Apprenticeship in thinking: Cognitive development in social context.* New York: Oxford University Press.

Sadler, Troy D. 2009. Situated learning in science education: Socio-scientific issues as contexts for practice. *Studies in Science Education* 45: 1–42.

Sampson, Victor, and Douglas Clark. 2008. Assessment of the ways students generate arguments in science education: Current perspectives and recommendations for future directions. *Science Education* 92: 447–472.

Savery, John R., and Thomas Duffy. 1995. Problem based learning: An instructional model and its constructivist framework. *Educational Technology* 35(5): 31–38.

Schwartz, Daniel L., John D. Bransford, and David Sears. 2005. Efficiency and innovation in transfer. In *Transfer of learning from a modern multidisciplinary perspective*, ed. Jose P. Mestre, 1–50. Greenwich: Information Age Publishing.

Scott, Philip H. 1998. Teacher talk and meaning making in science classrooms: A Vygotskian analysis and review. *Studies in Science Education* 32: 45–80.

Severiens, Sabine E., and Henk G. Schmidt. 2008. Academic and social integration and study progress in problem based learning. *Higher Education: The International Journal of Higher Education and Educational Planning* 58(1): 59–69.

Strobel, Johannes, and Angela van Barneveld. 2009. When is PBL more effective? A meta-synthesis of meta-analyses comparing PBL to conventional classrooms. *The Interdisciplinary Journal of Problem-Based Learning* 3(1): 44–58.

Szarlan, John, Suman Singha, and Scott W. Brown. 2010. *Striving for excellence: A manual for goal achievement*. Boulder: Pearson Publishing.

U.S. Department of Education, National Center for Education Statistics. 2014. Annual earnings of young adults. The condition of education 2014 (NCES 2014–083). Accessed January 8, 2015. http://nces.ed.gov/programs/coe/indicator_cba.asp.

Walker, Andrew, and Heather Leary. 2009. A problem based learning meta analysis: Differences across problem types, implementation types, disciplines, and assessment levels. *Interdisciplinary Journal of Problem-Based Learning* 3(1): 6–28.

Webb, Noreen M. 2010. Information processing approaches to collaborative learning. In *The international handbook of collaborative learning*, ed. Cindy E. Hmelo-Silver, Clark A. Chinn, Carol Chan, and Angela M. O'Donnell. New York: Routledge, Taylor & Francis.

Wenger, Etienne. 1998. *Communities of practice: Learning, meaning and identity*. Cambridge: Cambridge University Press.

Wertsch, James V. 1985. *Vygotsky and the social formation of mind*. Cambridge, MA: Harvard University Press.

Yukhymenko, Mariya, and Scott W. Brown. 2011. Motivation and prospective engagement related to global political processes: A cross-cultural analysis. Paper presented at the annual Association for Psychological Science, Washington, DC, May 2011.

Yukhymenko-Lescroart, Mariya, Scott W. Brown, Kimberly A. Lawless, Kamila Brodowinska, and Gregory P. Mullin. 2014. Thematic analysis of teacher instructional practices and student responses in middle school classrooms with problem-based learning environment. *Global Education Review* 1(3): 93–110.

Zepke, Nick, and Linda Leach. 2010. Improving student engagement: Ten proposals for action. *Active Learning in Higher Education* 11(3): 167–177.

Zimmerman, Barry J., and Albert Bandura. 1994. Impact of self-regulatory influences on writing course attainment. *American Educational Research Journal* 31: 845–862.

CHAPTER 5

"Out of Order!"
Exposure, Experience, E-Learning, and Evaluation: An Interdisciplinary Studies Approach @ Service Learning

Elaine Correa

Abstract The integration of interdisciplinary studies with service learning can invoke meaningful, engaging, and sustainable learning with technology for students beyond the classroom. Service learning provides learners with opportunities to explore the real needs of a community, and connect content knowledge with prior experiences from both the classroom and life. Grounded in Dewey's notion of "learning by doing," service learning necessitates deep reflection, as students merge theory with practice. An interdisciplinary studies approach to service learning offers space wherein the adage still applies for students today, "Tell me, and I will forget. Show me, and I may remember. Involve me, and I will understand."

Keywords Dewey • Freire • Experiences • Interdisciplinary approaches • Technology • Service

E. Correa (✉)
Division of Education, Medaille College, Buffalo, NY, USA

© The Author(s) 2016
R.D. Lansiquot (ed.), *Technology, Theory, and Practice in Interdisciplinary STEM Programs*,
DOI 10.1057/978-1-137-56739-0_5

Within the traditional system of education, there remains an "unwritten order" in which learning is somehow expected to occur. This "order" is often assumed to provide a legitimate sequencing for learning, since it aligns historically, and traditionally, with the accepted practices for teaching and imparting knowledge to students. Upon closer examination, however, it may be useful to challenge the order of our practices, particularly in terms of how some methods may impact exposure to material, through the intertwining of experiences that students have accumulated over time, as related to theory. Additionally, with the shift toward e-learning as a more integrated form that is utilized for teaching and evaluation, there are definitive justifications for reassessing what we do, and how we do it. The examination of exposure, experience, e-learning, and evaluation, as these elements are made evident in the interconnections that are driven by an interdisciplinary studies approach and manifested through service-learning projects, may help to shed some light on the value of learning out of order.

The Traditional Order of Learning

In the traditional classroom structure, the roles and expectations for students and instructors tend to follow a hierarchical structure of exchange, with the direction of learning organized by the instructor. From the perspective of the administration, faculty members in the USA are required to organize course learning objectives that align with specific content materials which comply with state-mandated regulations. These course learning objectives are expected to be conveyed to students in a systematic order or manner. Often with the use of a lecture format, sometimes enhanced with PowerPoint slides, faculty present material to students through this commonly accepted method for content delivery. In addition to the formal identification of information delivered by the instructor, the Socratic method of exchange may also accompany the lecture format as a means of engaging students in critical discussion. However, despite the benefits that the traditional method of instruction may offer faculty (whose intent is to impart knowledge to students), this approach to teaching is commonly viewed as disengaging to students and their learning.[1] Students often feel distanced from active participation in this structured learning approach to content delivery. Even with the enhancement of PowerPoint slides to the lecture format of teaching, students are subjected to a position of learning from the side-lines, watching rather than actively participating in their own learning experiences. According to Freire, this "banking concept of education" is ultimately counter-productive to learning. In this approach to education, students are expected to function like a "machine," as passive recipients that patiently receive, memorize, and repeat what is taught, after receiving, filing, and storing "deposits" of information."[2]

Ironically, although students today may be familiar with the technical uses of computer systems which enable them to receive documents, store large amounts of data, and file information on different networks, these deposits of information are functions which service students' needs for accessing information. These functions are not viewed as the sole part of the actual learning process. The computerized system of "receiving, storing, and filing" serve solely as a perfunctory system for "managing data." Like the "banking system of education" where teachers "deposit information" the computerized system of "data management" serves as a "repository of information" from which students can actively draw, revise, and modify the information that they keep.[3] The subjugation of students to learning that resembles functioning like a machine clearly relegates students to remain a "voiceless non-interactive presence" in their own learning. This passive position negates opportunities in which students can actively engage and assume responsibility for what and how they learn.

Furthermore, the impact of the banking' method of instruction, while possibly beneficial for some students, often results in limitations related to "context specific learning" and the "level of depth in application of the learning outcomes." A uni-directional form of instruction, through the use of lectures, is less productive for promoting thought or active engagement in problem-solving, which potentially can impact change in thinking or behavior.[4] As Doug Ward argues, lectures reinforce ideas, values and habits that students have already accepted. For example, while students may "learn" material through "drill and test" practices or "rote memory exercises," they often encounter great difficulty in application of their learning to other similarly situated contexts.[5] Additionally, the shallow nature of such learning reduces the depth of personal understanding and reflection that new information could potentially bring to their lives. Facts, theories, and concepts delivered in lectures, have little value, as Derek Bok contends, unless students can apply them to new situations, ask pertinent questions, make reasoned judgments, and arrive at meaningful conclusions.[6] Thus, a space must be created for students to serve as both knowledge consumers and producers instead of simply passive and receptive learners. An interdisciplinary approach which incorporates service learning opportunities may offer the type of vehicle needed for connecting knowledge with learning by challenging the order and structure inherent in the traditional format of teaching.

"Out of Order"
An Interdisciplinary Approach @ Service Learning

The integration of interdisciplinary studies with service learning can serve as an alternative pedagogical approach that invokes meaningful, engaging, and sustainable learning with technology for students beyond the classroom. However, in order to ensure that students are able to effectively transfer knowledge between courses, across disciplines, or research fields, the widespread adherence to traditional structures of teaching and learning must be continually challenged and revised. This necessitates replacing traditional pedagogical practices with constructivist theory and practice in order to reflect the educational needs of connecting technology to contemporary learning in interdisciplinary studies.

In a digitally based society, it should no longer be perceived as radical or out of order to move beyond textbooks and teacher-directed instruction toward more experiential forms of learning. Service learning provides learners with opportunities to explore the real needs of a community as well as to reflect and connect content knowledge with prior experiences from the classroom and life, while still fulfilling curriculum requirements.[7] Grounded in the notion of "learning by doing,"[8] service learning necessitates deep reflection in which students draw connections, merge theory with practice, and engage in critical discussions around contemporary educational issues. As constructivist theory recognizes, knowledge and reality are built inside the learner, as the importance of social interactions, collaboration, and connections are reinforced through the learning process. The practice of service learning supports constructivist epistemology of teaching and learning with heavy reliance on the reflective practices of inquiry and engagement.[9]

The benefits of an interdisciplinary studies approach to service learning highlights the impact of exposure, experience, e-learning, and evaluation, which are integral components for students to exercise control, responsibility, and ownership over their own learning. From the early conceptions of service learning that were directed toward students under the age of 18 in the USA in the early 1980s,[10] to the increased growth in designing service-learning programs with measurable benefits in the areas of academic achievement, social responsibility, and civic engagement,[11] it appears that the benefits of service learning have had significant impact at the elementary and secondary school levels.

At the university level, the shift toward greater student engagement also seems to be making some headway as aligned with teaching methodologies directed to specific program development. The educational research literature continues to recommend active student engagement in learning through a variety of approaches and methods that encourage student ownership and empowerment. Some of the types of activities that are deemed valuable for such engagement are the use of electronic and interactive media, encouraging more undergraduate level research, the structuring of collaborative learning experiences, and the developing of project-based learning.[12] Service learning is aligned with course bearing credits that deliberately focus attention on interactive integration of community service activities with educational objectives. Service learning is a teaching strategy "by which students learn and develop through active participation in thoughtfully organized service experience that meet actual community needs."[13] Thus, service learning combines the acquisition of content knowledge with reflective real-world experiences by supporting partnerships that benefit both the community and the student.

The concept of service learning is often aligned with the educational concepts of experiential learning or project-based learning. The project-based approach is anchored in the broader epistemological framework of constructivism,[14] with the premise that learners should be provided opportunities to construct their own meaning based out of their experiences of participating in a project. Additionally, working with peers should create multiple opportunities for meaningful learning to occur. Hence, by directly engaging the learner with content-related problems, educators are channeling student learning through these authentic learning experiences where learners discover a fact, concept, or principle on their own.

Experiential learning, such as service learning, is not new, and its roots can be traced back as far as the early 1900s with the writings of John Dewey. In *Experience in Education*, Dewey argued that "there is an intimate and necessary relationship between the processes of actual experience and education."[15] He believed that experience in itself is not education, but that the reflection upon this experience creates learning and meaning. Service learning provides students with real experience in the community and promotes reflection, discussion, and critical thinking

in the classroom.[16] The process of reflection, discussion, and critical thinking is paramount for a successful service-learning program. Ernest Boyer[17]
points out that the opportunities that arise from service-learning can foster a sense of service; this sense needs to be instilled at a young age, well
before students begin college. Although many schools have incorporated
service learning opportunities for students at the elementary and high
school social studies program level, there is a real need for more active
service learning engagement at the college level.

Service Learning and Student Learning

The premise of service learning is that students participate in society and
inherently internalize the values of the community which, in turn, they
promote and pass onward to the next generation. As students actively participate in the community, they are seen as valuable contributors to society, aligned with active participation instead of reinforcing a focus on their
shortcomings.[18] With the election and re-election of President Obama,
there has been a renewed call for service in higher education. In his inaugural speech, the President stated,

> What is required of us now is a new era of responsibility—recognition, on
> the part of every American, that we have duties to ourselves, our nation, and
> the world, duties that we do not grudgingly accept but rather seize gladly,
> firm in the knowledge that there is nothing so satisfying to the spirit, so
> defining of our character, than giving our all to a difficult task.[19]

The chief proponent of service learning as an educational strategy
focuses on guiding learners toward exploring and connecting knowledge
with real-life experiences. Derived from the principles of engaged learning of William James, John Dewey, James Coleman, Ernest Boyer, Paulo
Freire,[20] and many others, the use of challenging but unique, and often
times innovative, learning experiences can foster greater involvement of
young adults in service, civic problem-solving, and community leadership.
In conjunction with academic learning outcomes, these opportunities
for authentic engagement provide students with meaningful experiences
where theory can be utilized in practice.

As many advocates of postmodern pedagogy or critical pedagogy con-
tend, reflective relationships between the teacher and the student, where
the teacher does not ask the student to accept the teacher's authority but

rather is invited to "join with the teacher in inquiry, into that which the student is experiencing,"[21] opens the door for exchange by challenging the prescribed order for learning. These uncensored or informally structured forms of communication can shape the meaning, interactions, and understanding of all participants invested in the educational enterprise. The teacher is not the model, the master, or the expert, but rather one contributor who assists in facilitating the learning experiences that extend beyond the classroom walls. As Doll has stated,

> I believe a new sense of educational order will emerge, as well as new relations between teachers and students, culminating in a new concept of curriculum. The linear, sequential, easily quantifiable ordering system dominating education today—one focusing on clear beginnings and definite endings—could give way to a more complex, pluralistic, unpredictable system or network.[22]

Thus, the envisioning of learning out of order through service learning opportunities necessitates a conscious shift in thinking about learning, and accepting new ways to brokering how learning occurs outside the traditional boundaries of exchange and interaction contained within the classroom walls. This paradigm shift must be internalized beyond the privileging of knowledge as rooted in dominant culture and language as reflected in current Western views of learning. As Anzaldua has insightfully argued, we should "Shift out of habitual formations; from convergent thinking, analytical reasoning that tends to use rationality to move toward a single goal (a Western mode), to divergent thinking, characterized by movement away from set patterns and goals and toward a more whole perspective, one that includes rather than excludes."[23] To move toward this type of shift from convergent to divergent thinking necessitates that other "funds of knowledge"[24] are not only viewed as valid, but valuable. As such, an assets-oriented approach recognizes that homegrown knowledge emanates from people in diverse locations and positions in society. Freire points out that critical pedagogy, communication, and dialogue are central for human understanding. [25] Because "social problems are social constructions,"[26] the opportunity whereby students communicate, engage, and exchange actively with people within the community as a component of service learning reflects the potential that these forms of interaction possess for generating greater understanding of the learning articulated within theoretical classroom discussions.

As students begin to examine these social constructions and the process of how the social problem manifests itself and becomes nameable and describable, they are able to critically analyze the process or processes that contribute to how the issue or problem emerged, and the results that arise from the social problem. Dialogue in social relations can in any context be oriented toward inclusion of different social agents, or conversely toward the imposition and/or inhibition of the rules of interaction. The construction and reconstruction of meaning is highly dependent on the orientation of the interaction. For example, vertical interaction maintains social control, promotes adaptation of the dominant patterns, and leads to standardization and conformity; conversely, horizontal communication sets the individuals on an equal plane and serves as a tool for social change or innovation.[27]

Service learning activities serve as a means to deconstruct the hegemony of the standard teaching practices embedded in the traditional banking system of education,[28] as learning and subsequent knowing are situated in the path of subsequent experience, not in the path of banking knowledge. It is important, however to recognize that experience in itself does not necessarily result in learning. As Dewey acknowledges, experiences can be "miseducative" and "educative." [29] Experiences categorized as educative for students occur when student's reflective thoughts create new meaning, leading to growth and informed actions. In connection with the premise of service learning, the definition of Dewey's concept of learning as educative relies on the following: (1) generating interest in the learner; (2) organizing learning that is intrinsically worthwhile to the learner; (3) ensuring that learners are presented with problems that awaken new curiosity and create a demand for information; and (4) enabling that a considerable time span is covered which can foster development of learning over time.[30] Service learning opportunities in courses that are structured to meet these four conditions extend the walls of the classroom into the community.

Stanton, Giles, and Cruz [31] argue that service learning is an effective model for creating and studying a highly authentic learning environment where instruction emphasizes the idea that a significant portion of what is learned is specific to the situations in which it is learned.[32] These experiences include activities that are participatory, interactive, and representative of real-world events,[33] and thus require students to actively problem-solve, think critically, and reflect on their learning.

"In-Order" Versus "Out of Order":
Critiques of Service Learning in Academe

Over the last decade, service learning has gained popularity in higher education and has been more readily embraced as a "scholarship of engagement"[34] that includes the scholarship of teaching and learning movement, experiential education, community-based research, and undergraduate research. Service learning is increasingly seen as an instrumental component in sustaining stronger relationships with local communities. Although there has been increased recognition of service learning in over 450 colleges and universities across the USA,[35] a number of critiques have emerged that question the established order of teaching with service learning in higher education. Undeniably, it should be anticipated that shifting fundamental teaching and learning paradigms which disrupt the conventional order of traditional practices or the status quo will invoke significant backlash. Similar to the critical discussions that arise as to the value of interdisciplinary approaches to teaching and learning, the critique directed toward service learning not only questions the substantive value this approach offers, but challenges this type of learning as nothing more than another educational fad that panders to the current rhetoric for educational change.

Critics often view service learning as a watering down, diluting, or weakening of the curriculum at the expense of quality in higher education. Furthermore, some critiques have argued that the time allocated for students to participate in service learning opportunities significantly detracts from opportunities where students could more productively utilize their time, such as studying in the library or testing results in the laboratory.[36] These concerns can be quickly addressed given the importance that is placed on the intentional selection of placements that must meet the learning objectives of a course that integrates service learning opportunities. Service learning partnerships should be organized to align with classroom activities that offer students opportunities to critically reflect on the nature of their experiences in relation to theoretical classroom learning. As Howard has argued, it is essential that opportunities are relevant and meaningful to both the community and the students. [37] Hence purposeful planning, design, and implementation of activities aligned to specific learning objectives must be organized in collaboration with the community benefiting from the service to avoid any conflicts between instructional and service objectives. Additionally, Butin argues, "Service-learning has immense transformational potential as a sustained, immersive, and consequential pedagogical practice."[38] His claim is partially supported by the fact that service learning has become a major presence within higher education, with a substantial number of faculty members across an increasingly diverse range of academic courses devoted to promoting its use.[39]

However, on-going critical debate continues to focus on the implicit power differential that are embedded in the duality of the service learning opportunity, where the binary of "those in need" are juxtaposed to "those providing help." Here, the notion of "oppressors" (those with power) working with "the oppressed" (those without power) reinstitutes a binary dynamic whereby the weak are subjugated to the visions of the strong and are thus expected to conform to the visions of the status quo, as ascribed through the service project. In essence, such critiques challenge the ability of students to understand the problem without reinstating the status quo, preventing them from identifying solutions and determining what is best for the populations they will serve. Furthermore, these critiques question the efficaciousness and authority of students to be able to respond to the problems encountered by those individuals who are viewed as very different from them.

The implications for service learning are clear, as the model suggests the importance of reflecting not only on the particular contexts of service learning activities but personal and ideological matters as well. For example, Butin rebuts a common misconception of critics of service learning: "The overarching assumption is that the students doing the service-learning are White, sheltered, middle-class, single, without children, un-indebted, and between ages 18 and 24. But that is not the demographics of higher education today, and it will be even less so in 20 years."[40] Critiques that are premised on "difference" as a means to obstruct or reduce understanding reflect a reductionist view and a limited understanding of how learning through dialogue and exchange can occur. Service learning is premised on fostering "border-crossings" across categories of race, ethnicity, class, (im)migrant status, language, sexuality, and (dis)ability. Furthermore, the assumption that students who participate in these opportunities will not be reflected in the populations that are served is premised on an expectation of a homogenized contemporary student population. In most cases, the student demographic that will participate in service learning opportunities consists of students who already occupy and cross a number of the socially constructed categories of race, class, ethnicity, (im)migrant status, language, sexuality and (dis)ability, as is currently reflected in the diverse student populations in higher education today.

Manchester and Baiocchi question whether service learning is more harmful than beneficial, particularly when the students' (and their instructors') motive to serve may supersede the actual needs of the community as defined by the community being served.[41] They raise the concern of whether the emphasis of attention should be placed on the instruction or on the service to the community. Implicit in their challenge is whether it is really possible to meet both the intended learning objectives and the community's needs at the same time. In response to this critique, it is imperative to recognize that the research indicates the mutually beneficial outcomes of well-implemented service learning programs that support the long-term development of civic responsibility, while at the same time, recognize the impact of activities that respond to the immediate intellectual and social benefits for students and the communities that they serve.[42]

Exposure, Experience, E-Learning, Evaluation

Despite the technological revolution that has entered the classroom and infiltrated various aspects of teaching, the expectation of how learning should occur continues to challenge faculty and students alike with the more recent debates focused on technological integration and student learning. In the past decade, the "digital divide" and tensions between the students who comprise the "digital generation," commonly referred to as "digital natives,"[43] and those born without the "digital DNA" or "digital footprint," known as "digital immigrants," have reflected divergent levels of integration of new media technologies into their respective lives. The debate of the relationship among digital technology, student participation, and learning trends has been argued extensively in the recent review of the literature,[44] and remains a contested space, vacillating attention between the limitations arising from the fear of an overly optimistic or pessimistic technological determinism that is accompanied by a dearth of empirical evidence.

With the widespread dissemination of the Internet and new media technology tools, a blurring of the lines between interactivity discourses of technology in learning and traditionalist approaches to teaching has become apparent. Advocates who support a traditional view and approach to education, focusing on reading, writing, and arithmetic ("the 3 Rs") may also recognize the benefits that some level of technological integration offers instruction and student learning. Additionally, current policy initiatives in the USA (as exemplified by the *New Media Consortium Horizons Report* and the Common Core States Standards Initiative[45]) have clearly identified the contemporary focus on greater technological integration in teaching, which has been simultaneously supported with the rapid physical restructuring and modifications to classrooms (from blackboards to interactive boards) that have occurred within schools and institutions of higher education. Furthermore, this massive swing toward supporting technological integration is evident in the ways that information and communication technology (ICT) are promoted in training and professional development initiatives targeting educators and faculty at all levels.

As previously discussed, the adherence to the traditional structure of teaching has made changing the format of classes more challenging for all stakeholders. The shift from a teacher-centered to a student-centered pedagogy has not occurred without resistance from teachers and students alike. Students have become accustomed to sitting passively in lectures, reviewing instructors' notes or slides that are posted online, physically attending lectures in a passive mode of engagement, and cramming for exams. In many ways, students have accepted and internalized this mode of learning by complying with the established protocol inherent in the banking education model.[46] As a consequence, students may have expectations and possible resentment toward faculty who challenge the traditional structures of teaching, and learning in which exposure, experience, and evaluation are not neatly defined. There is indeed comfort in following the known format for learning where the professor is responsible as the agent for active exchange and the student is only expected to serve as the passive recipient. In this framework, students may be more prone toward developing a deep fear of failure and for taking intellectual risks. There are benefits from clear, firm solutions to academic problems that do not require students to push beyond a single "right" answer. Encouraging students to embrace the change in learning order may initially be intimidating and destabilizing as the rules of the game are not so neatly organized, and the consequences for learning in this way may not be viewed as visibly straightforward.

The attention to evaluation and grade assessment is perhaps one of the most basic concerns that students identify when thinking about service learning. "*What grade will I receive?* and *How will this impact my grade?*" are familiar concerns echoed by students in their first service learning experience. The evaluation of service learning is similar to other forms of active learning. Students must understand how the learning objectives are matched to the service opportunity, and they must understand their role as well as the expectations that are required for evaluation and assessment purposes. As service learning opportunities provide students with real-world problems, a great deal of planning based on the appropriate skills (theoretical and practical) as developed and discussed in the classroom become part of the way students begin to analyze the service project. Students are expected to devise a plan or blue-print based on their analysis of the problem and their ideas in responding with possible solutions. The evaluation of the student's knowledge and critical thinking abilities become an integral part of the assessment, and this may also serve as part of the grade earned for the service learning project. Assessment and evaluation processes vary with the different needs being met with the service learning project. The alignment of learning objectives with service requirements provides a starting point from where evaluation criteria are identified in relation to grade assessment.

From Classroom to Community to Cyberspace: E-Learning @ Service Learning

National studies support the value of service learning programs in extending learning beyond simple delivery toward active engagement in the exchange of ideas and experiences. In higher education, the benefits of authentic service learning opportunities result in meaningful, challenging experiences that encourage problem-solving within a real-world context in which skills of critical thinking, collaboration, and community engagement intersect and generate possible solutions within a specific timeframe. As social interactions play a fundamental role in service learning activities, where students are expected to engage in discussion, ask questions, state opinions, negotiate meanings and resolve conflicts, critics of service learning often question whether technology is needed for social interaction. How can the integration of technology in learning support real service learning partnerships? Returning back to the work of Dewey, communication, particularly face-to-face discourse, is considered integral to creating meaningful educative experiences.[47] Accordingly in a community where full and open communication exists, one finds an essential condition for values that inform behavior. Within the framework of technological integration, how might exposure and experience connect to e-learning and subsequent evaluation if face-to-face communication is integral to under-standing the lived experiences of the community that students will serve?

Moving from Theory to Practice: Course, Semester, Year

Service learning opportunities begin with initial interaction between the service provider and students. For students to really understand the needs of the service partner, they must spend time observing, participating, and interacting with the various constituents within the specific service project. ICT-based civic participation in education is an increasingly important element of the recommendations and interventions addressing both the formal setting of learning, such as schools and higher education, as well as informal learning environments, such as local communities.[48] Educational researchers have provided models and strategies to explore both the pitfalls and the potential of creating a technology-integrated project-based learning environment. Technologically integrated projects do not automatically mean that there are no face-to-face interactions between the students and the service provider. Rather, technology is used as another means to connect with the service provider utilizing the expertise of student's digital skills and interests. It is important to keep in mind that service learning helps students with the interviewing process, increases students' confidence in dealing with people, and enhances their team work communication skills.[49]

The mastering of academic content standards, while immersed in hands-on, technology-integrated projects which provide learning experiences that are not usually possible within the confines of the traditional classroom, requires significant planning. The use of technology in specific service learning opportunities is one way to harness some of the skills and competencies of students in developing appropriate responses to the needs of the service learning partners in the community. As one of the models that can be utilized, technological expertise provides an appropriate venue in helping students contribute back to the community using the skills they already possess (i.e., simply by their own interest with technology). As there are numerous types of service learning projects, the steps that are involved in supporting service initiatives with the use of technology will be discussed within the framework of three different levels: the course, the semester, and the academic year.

One specific way in which technological integration can be utilized in a service learning project is in the development of media vignettes for partner institutions to use with their clients. For example, in a Psychology course with a focus on patient relations with community services, students were offered a service learning project to create appropriate digital media to assist providers in reaching the various needs of their clients. This type of project required that students visit the site, speak with the staff and clients, and observe the routines and interactions of the community. This type of project would require a full semester to complete, as time would be needed for students to visit and interview members of the staff who work with the clients, as well as devote time to returning to class to discuss their observations before generating a blue-print for their service project plan. In groups, this type of service learning project can generate collaborative working relations between students and the staff, as well as provide diverse perspectives in the assessment of the needs of the client as articulated in the solution plan. Because these types of team service projects require off-campus visits to the site, a great deal of organizational work and planning needs to be completed prior to the start of the academic semester. As such, these types of service initiatives must be organized with contingency options if changes to the funding of providers or their needs should occur prior to the start of the semester.

These types of opportunities can be very enriching for students, facilitating interpersonal skills and encouraging the need for dialogue, as Eyler, Giles, Stenson, and Gray have found.[50] The outcome of this type of full semester project can result in students feeling a sense of accomplishment for contributing back to the community through their service. Additionally, students may experience a sense of pride in realizing that their digital savvy skills were used to produce or update websites, create videos, or simply design/organize a framework for categorizing the literature from the provider in assisting the needs of clients, by being more accessible and user-friendly.

Service learning projects that require the commitment of several months of student engagement may be difficult for most faculty members to embrace and incorporate over the course of a semester. Thus, shorter time-restricted service learning projects may be easier to organize in classes to accompany specific learning objectives. In general or liberal studies courses, students might be asked to participate in a clothing or food drive. These types of service learning projects can be used for students to work in large groups to address a local community concern using their knowledge of theories that have been discussed in the classroom. In these specific types of service projects, students might find the use of technology extremely engaging for

learning more about the project (from the client website), to actually working on the project using technology. For example, students might use technology to advertise the project, increase attention and interest of their peers
through social media, elicit participation of peers, and generally discuss the
project with classmates. One alternative way to promote and encourage participation in service learning initiatives is to connect the outcomes of the
project with the interests of students. In these shorter service learning initiatives, students can demonstrate their technological skills by creating on-line
evaluations for assessing the impact of their project on the campus community. Here the use of on-line polls and/or surveys, as well as creating an
assessment tool for evaluating the success of the class to complete the project could be another means for technological integration. The large group
project-based approach offers students an experimental, interactive, investigative and cooperative form of learning that opens a space for students to
actively engage in their own learning through the use of technology.

For senior level courses which focus on the writing of a final thesis or
cumulative research paper, the use of a technologically integrated service
learning project may assist in providing students with an innovative means
for merging theory with practice. Students may utilize technology as a
means to create a response to an issue that demonstrates their ability to
apply new concepts in complex, meaningful ways. With the use of technology in a service learning project that extends the full academic year,
students can actively participate in an experimental form of learning which
is interactive, investigative, and cooperative. For example, students in
Mathematics or Business could work with a service partner in assessing
the effectiveness of the organization to meet the needs of their clients.
This type of full year service learning project would be akin to a research
study whereby the data for the paper would be collected by the students
through a review of the organizational structure. With the use of technology, students would review the bookkeeping/accounting of the service
partner and provide a cost/benefits ratio analysis of the services offered.
Recommendations for future directions would be identified and both the
service provider and faculty member would evaluate the project after an
oral presentation by the student. Much of the work involved in this type
of service project would be on-line, but the final product would reflect the
communication skills of the student in working with diverse populations
in the community. The project would be divided into three phases of planning, creating, and processing,[51] with each phase consisting of collaborative
learning and opportunities for processing information, reflecting, interro-
gating, and integrating new information into the students' pre-existing

knowledge. Through this type of engagement, students can become skilled at developing evidence-based arguments by discovering facts, concepts, and principles through their own interactions with the service provider.

CONCLUSION

Through experiential and project-based learning, students receive diverse opportunities to develop different skills and contribute to projects that will serve the needs of specific communities. Inherent in this model is a requirement that students must work collaboratively to determine what impact they will have as reflected by the tool or piece that they create. Additionally, students are expected to identify how they will harness the expertise they need through the development of a written plan/project proposal. The objective for documenting their plan/project proposal is to ensure that the direction the students assume will be meaningful to the community they are servicing. Furthermore, students must create an assessment/evaluation tool of the project piece to demonstrate their effectiveness and impact in addressing the needs of their community partner. Due to its connection to various forms of technology, its educational efficacy, and impact, a service learning approach within an interdisciplinary framework should have profound value to active student learning. As the great Confucian philosopher of the classical period, Xunzi, said, "Tell me, and I will forget. Show me, and I may remember. Involve me, and I will understand."[52] Ultimately, an interdisciplinary approach to service learning offers the space wherein this adage of the past holds true for students today.

NOTES

1. Bell Hooks, *Teaching Community: A Pedagogy of Hope* (New York: Routledge, 2003).
2. Paulo Friere, *Pedagogy of the Oppressed*, trans. by Myra Bergman Ramos (New York: Seabury Press, 1970).
3. Ibid.
4. Doug Ward, "Why Change Our Teaching?" *Faculty Focus*, reprint from *Teaching Matters*, July 15, 2015.
5. Ibid.
6. Derek Bok, *Our Underachieving Colleges: A Candid Look at How Much Students Learn and Why They Should Be Learning More* (New Jersey: Princeton University Press, 2007).

7. Jennifer Avenatti, Jennifer Twila D. Garza, Ambrose P. Panico, and Students, "The Adventure of Service." *Reclaiming Children and Youth* 16, no. 1 (2007): 28–32.

8. John Dewey, *Experience in Education* (New York: Macmillan, 1939).

9. Martin Dougiams, *Moodle: Open-Source Software for Producing Internet-Based Courses*, 2001, http://moodle.com/.

10. Daniel Conrad and Diane Hedin, "School-Based Community Service: What We Know from Research and Theory" *Phi Delta Kappan* 72, no. 10 (1991): 743–49; James Kielsmeir, "Build a Bridge Between Service and Learning," *Phi Delta Kappan* 91, no. 5 (2010): 8–15; Rebecca Skinner and Christopher Chapman, *Service-learning and Community Service in K–12 Public Schools* (Washington, DC: National Center for Education Statistics, 1999) http://nces.ed.gov/surveys/frss/publications/1999043/index. asp; Karen Pittman. "Balancing the Equation: Communities Supporting Youth, Youth Supporting Communities," *Community Youth Development (CYD) Anthropology* (2002): 19–24.

11. Shelley Billig, Sue Root, and Dan Jesse, "The Relationship between Quality Indicators of Service-Learning and Student Outcomes: Testing the Professional Wisdom," in *Advances in Service Learning Research: Vol. 5. Improving Service-Learning Practice: Research on Models That Enhance Impacts*, ed. Sue Root, Jane Callahan and Shelley Billig, 97–115 (Greenwich, CT: Information Age, 2005); Carnegie Corporation and CIRCLE, *The Civic Mission of Schools* (New York: Carnegie Corporation of New York, 2003); Daniel Conrad and Diane Hedin, "School-Based Community Service: What We Know from Research and Theory," *Phi Delta Kappan* 72, no. 10 (1991): 743–49; National Commission on Service Learning Executive Summary: Learning in *Deed: The Power of Service Learning for American Schools* (Newton, MA: National Commission on Service Learning, 2002), accessed June 15, 2015, http://learningindeed.org.

12. Robert Bringle and Julie Hatcher, "Reflection in Service Learning: Making Meaning of Experience," *Educational Horizons* Summer (1999): 179–85.

13. Jennifer Avenatti, Twila D. Garza, Ambrose P. Panico, and Students, "The Adventure of Service," *Reclaiming Children and Youth* 16, no. 1 (2007): 28–32.

14. Jean Piaget, *The Equilibrium of Cognitive Structures: The Central Problem of Intellectual Development* (Chicago: University of Chicago Press, 1985).
15. John Dewey, *Experience in Education* (New York: Macmillan, 1939), 6.
16. Jeffery Moser and George Rogers, "The Power of Linking Service to Learning," *Tech Directions* 64, no. 7 (2005): 18–21.
17. Ernest Boyer, *High School: A Report on Secondary Education in America* (New York: Harper & Row, 1983).
18. James Kielsmeier, *Growing to Greatness 2005: The State of Service Learning Project* (Saint Paul: National Youth Leadership Council, 2005).
19. Barack Obama, Inaugural Address 2009, accessed June 15, 2015, https://www.whitehouse.gov/blog/inaugural-address/.
20. William James, *The Moral Equivalent of War: International Conciliation, No. 27, February, 1910* (Whitefish, MT: Literary Licensing, 2013); Dewey, *Experience in Education*; James Coleman, "Differences Between Experiential and Classroom Learning," *Experiential Learning: Rationale, Characteristics, and Assessments*, ed. Morris T. Keeton and Associates, 46–61 (San Francisco: Jossey-Bass, 1977); Ernest Boyer, "The Scholarship of Engagement." *Journal of Public Service and Outreach* 1, no. 1 (1996): 11–20; Paulo Friere, *Pedagogy of the Oppressed*, trans. by Myra Bergman Ramos (New York: Seabury Press, 1970).
21. William Doll, *A Post-Modern Perspective on Curriculum* (New York: Teachers College Press, 1993).
22. Ibid, 3.
23. Gloria Anzaldua, *Borderlands/La Frontera: The New Mestiza*, 2nd ed. (San Francisco: Aunt Lute Foundations Books, 1999).
24. Luis, C. Moll and Norma Gonzalez, "Engaging Life: A Funds of Knowledge Approach to Multicultural Education," in *Handbook of Research on Multicultural Education*, 2nd ed, ed. James Banks and Cherry McGee Banks, 669–715 (New York: Jossey-Bass, 2004).
25. Friere, *Pedagogy of the Oppressed*.
26. Michel Foucault, *The Archaeology of Knowledge*, trans. by T.A.M. Sheridan Smith (New York: Pantheon Books, 1972).
27. Serge Moscovici, "Social Influence and Conformity," in *Handbook of Social Psychology*, 2, ed. Gardner Lindzey and Elliot Aronson, 347–412 (New York: McGraw-Hill, 1985).

28. Friere, *Pedagogy of the Oppressed.*
29. Dewey, *Experience in Education.*
30. Ibid.
31. Timothy Stanton, Dwight Giles, and Nadinne Cruz, *Service-learning: A Movement's Pioneers Reflect on Its Origins, Practice, and Future* (San Francisco: Jossey Bass Publishers, 1999).
32. Aldrin E. Sweeney and Jeffery A. Paradis, "Developing a Laboratory Model for the Professional Preparation of Future Science Teachers: A Situated Cognition Perspective," *Research in Science Education* 34, no. (2004): 195–219.
33. Etienne Wenger, Richard McDermott, and William M. Snyder, *A Guide to Managing Knowledge: Cultivating Communities of Practice* (Cambridge, MA: Harvard University Press, 2002).
34. Ernest Boyer, "The Scholarship of Engagement," *Journal of Public Service and Outreach* 1, no. 1 (1996): 11–20; Lee Shulman, *Teaching as Community Property* (San Francisco: Jossey-Bass, 2004).
35. Dan Butin, "The Limits of Service-Learning in Higher Education," *Review of Higher Education* 29, no. 4 (2006): 473–98.
36. Maryann Gray, Elizabeth Ondaatje, and Laura Zakaras, *Combining Service and Learning in Higher Education. Summary Report* (Santa Monica, CA: Rand, 1999).
37. Jeffery P. F. Howard, "Academic Service Learning: A Counter-Normative Pedagogy," in *Academic Service Learning: A Pedagogy of Action and Reflection*, ed. Robert Rhoads and Jeffery Howard, 21–9 (San Francisco: Jossey-Bass Publishers, 1998).
38. Butin, "The Limits of Service-Learning in Higher Education."
39. Ibid.
40. Ibid.
41. Heather Manchester and Lauren Baiocchi, "Reflecting on Jonathan Kozol's Challenge: Who Really Benefits from Service Learning?" *Currents: The Newsletter of Youth Service California* IX, no. 3 (2001), http://www.energizeinc.com/art/vvser.html (accessed June 15, 2015).
42. Shelley Billig and Andrew Furco, *Service-Learning Through a Multidisciplinary Lens: Advances in Service Learning Research* (Greenwich, CT: Information Age, 2002); Meta Mendel-Reyes, "A Pedagogy for Citizenship: Service Learning and Democratic Education," in *Academic Service Learning: A Pedagogy of Action*

and Reflection, ed. Robert Rhoads and Jeffery Howard, 31–8 (San Francisco: Jossey-Bass Publishers, 1998).

43. Marc Prensky, *Teaching Digital Natives: Partnering for Real Learning* (California: Corwin, 2010).

44. David Buckingham, "Beyond Technology: Children's Learning in the Age of Digital Culture (Cambridge: Polity, 2007); Lance Bennett, Changing Citizenship in the Digital Age", in *Civic Life Online: Learning How Digital Media Can Engage Youth*, ed. Lance W. Bennett (Cambridge, MA: MIT Press, 2008).

45. Larry Johnson, Samantha Adams Becker, Victoria Estrada, and Alex Freeman, *NMC Horizon Report: 2014 Higher Education Edition* (Austin, TX: The New Media Consortium, 2014); Council of Chief State School Officers (CCSSO) and National Governors Association Center for Best Practices (NGA Center), "Common Core State Standards Preparing America's students for College and Career," 2009, http://www.corestandards.org/about-the-standards/.

46. Friere, *Pedagogy of the Oppressed*.

47. Dewey, *Experience in Education*.

48. Mark Warschauer, *Technology and Social Inclusion: Rethinking the Digital Divide* (Cambridge, MA: MIT Press, 2003); Neil Selwyn, *Citizenship, Technology and Learning: A Review of Recent Literature* (Bristol, UK: Futurelab, 2007), accessed June 15, 2015, http://www.futurelab.org.uk/resources/documents/lit_reviews/Citizenship_Review_update.pdf.

49. Mary Tucker, Anne McCarthy, John Hoxmeier, and Margarite Lenk, "Service-Learning Increases Communication Skills across the Business Curriculum," *Business Communication Quarterly* 61, no. 2 (1998): 89–100.

50. Janet Eyler, Dwight Giles, Christine Stenson, and Charlene Gray. *At a Glance: What We Know About the Effects of Service-Learning on College Students, Faculty, Institutions and Communities, 1993–2000* (Washington, DC: Learn and Serve America National Service Learning Clearinghouse, 2001), accessed June 15, 2015, http://servicelearning.org.

51. Lillian Katz and Sylvia Chard, *Engaging Children's Minds: The Project Approach* (Connecticut: Ablex, 2000).

52. *Confucius: Analects—With Selections from Traditional Commentaries*, trans. Edward Slingerland (Indianapolis: Hackett Publishing). The original work was published c. 551–450 BCE.

BIBLIOGRAPHY

Anzaldua, Gloria. 1999. *Borderlands/La Frontera: The new mestiza*, 2nd ed. San Francisco: Aunt Lute Foundations Books.

Avenatti, Jennifer, Twila D. Garza, Ambrose P. Panico, and Students. 2007. The adventure of service. *Reclaiming Children and Youth* 16(1): 28–32.

Bennett, Lance W. 2008. Changing citizenship in the digital age. In *Civic life online: Learning how digital media can engage youth*, ed. Lance W. Bennett, 1–24. Cambridge, MA: MIT Press.

Billig, Shelley, and Andrew Furco. 2002. *Service-learning through a multidisciplinary lens: Advances in service learning research*. Greenwich: Information Age.

Billig, Shelley, Sue Root, and Dan Jesse. 2005. The relationship between quality indicators of service-learning and student outcomes: Testing the professional wisdom. In *Advances in service learning research: Vol. 5. Improving service-learning practice: Research on models that enhance impacts*, ed. Sue Root, Jane Callahan, and Shelley Billig, 97–115. Greenwich: Information Age.

Bok, Derek. 2007. *Our underachieving colleges: A candid look at how much students learn and why they should be learning more*. Princeton: Princeton University Press.

Boyer, Ernest. 1983. *High school: A report on secondary education in America*. New York: Harper & Row.

Boyer, Ernest. 1996. The scholarship of engagement. *Journal of Public Service and Outreach* 1(1): 11–20.

Bringle, Robert, and Julie Hatcher. 1999. Reflection in service learning: Making meaning of experience. *Educational Horizons Summer* 77: 179–185.

Buckingham, David. 2007. *Beyond technology: Children's learning in the age of digital culture*. Cambridge: Polity.

Buckingham, David. 2010. Minding the gaps: Teachers' cultures, students' cultures. In *Adolescents' online literacies: Connecting classrooms, media and paradigms*, ed. Donna Alvermann, 183–203. New York: Peter Lang.

Butin, Dan. 2006. The limits of service-learning in higher education. *The Review of Higher Education* 29(4): 473–498.

Carnegie Corporation and CIRCLE. 2003. *The civic mission of schools*. New York: Carnegie Corporation of New York.

Coleman, James. 1977. Differences between experiential and classroom learning. In *Experiential learning: Rationale, characteristics, and assessments*, ed. Morris T. Keeton, and Associates, 46–61. San Francisco: Jossey-Bass.

Confucius: Analects—With Selections from Traditional Commentaries. 2003. Trans. Edward Slingerland. Indianapolis: Hackett Publishing.

Conrad, Daniel, and Diane Hedin. 1991. School-based community service: What we know from research and theory. *Phi Delta Kappan* 72(10): 743–749.

Council of Chief State School Officers (CCSSO) and National Governors Association Center for Best Practices (NGA Center). 2009. Common core state standards preparing America's students for college and career. Accessed June 15, 2015. http://www.corestandards.org/about-the-standards/.

Dewey, John. 1939. *Experience in education*. New York: Macmillan.

Doll, William. 1993. *A post-modern perspective on curriculum*. New York: Teachers College Press.

Dougiamas, Martin. 2001. Moodle: Open-source software for producing Internet-based courses. http://moodle.com.

Eyler, Janet, Dwight Giles, Christine Stenson, and Charlene Gray. 2001. *At a glance: What we know about the effects of service-learning on college students, faculty, institutions and communities, 1993–2000*. Washington, DC: Learn and Serve America National Service Learning Clearinghouse. Accessed June 15, 2015. http://servicelearning.org.

Foucault, Michel. 1972. *The Archaeology of Knowledge*. Trans. T.A.M. Sheridan Smith. New York: Pantheon Books.

Freire, Paulo. 1970. *Pedagogy of the Oppressed*. Trans. Myra Bergman Ramos. New York: Seabury Press.

Gray, Maryann, Elizabeth Ondaatje, and Laura Zakaras. 1999. *Combining service and learning in higher education. Summary report*. Santa Monica: Rand.

Hooks, Bell. 2003. *Teaching community: A pedagogy of hope*. New York: Routledge.

Howard, Jeffery P.F. 1998. Academic service learning: A counter-normative pedagogy. In *Academic service learning: A pedagogy of action and reflection*, ed. Robert Rhoads and Jeffery Howard, 21–29. San Francisco: Jossey-Bass Publishers.

James, William. 2013. *The moral equivalent of war: International conciliation, No. 27, February, 1910*. Whitefish: Literary Licensing.

Johnson, Larry, Samantha Adams Becker, Victoria Estrada, and Alex Freeman. 2014. *NMC Horizon Report: 2014 higher education edition*. Austin: The New Media Consortium.

Katz, Lillian, and Sylvia Chard. 2000. *Engaging children's minds: The project approach*. Connecticut: Ablex.

Kielsmeier, James. 2005. *Growing to greatness 2005: The state of service learning project*. Saint Paul: National Youth Leadership Council.

Kielsmeier, James. 2010. Build a bridge between service and learning. *Phi Delta Kappan* 91(5): 8–15.

Lee, Shulman. 2004. *Teaching as community property*. San Francisco: Jossey-Bass.

Manchester, Heather, and Lauren Baiocchi. 2001. Reflecting on Jonathan Kozol's challenge: Who really benefits from service learning? *Currents: The Newsletter of Youth Service California* IX(3). Accessed June 15, 2015. www.energizeinc.com/art/vvser.html.

Mendel-Reyes, Meta. 1998. A pedagogy for citizenship: Service learning and democratic education. In *Academic service learning: A pedagogy of action and reflection*, ed. Robert Rhoads and Jeffery Howard, 31–38. San Francisco: Jossey-Bass.

Moll, Luis, and Norma Gonzalez. 2004. Engaging life: A funds of knowledge approach to multicultural education. In *Handbook of research on multicultural education*, 2nd ed, ed. James Banks and Cherry McGee Banks, 669–715. New York: Jossey-Bass.

Moscovici, Serge. 1985. Social influence and conformity. In *Handbook of social psychology*, vol. 2, ed. Gardner Lindzey and Elliot Aronson, 347–412. New York: McGraw-Hill.

Moser, Jeffery, and George Rogers. 2005. The power of linking service to learning. *Tech Directions* 64(7): 18–21.

National Commission on Service Learning. Executive summary: Learning. 2002. In *Deed: The power of service learning for American schools*. Newton: National Commission on Service Learning. Accessed June 15, 2015. http://learningindeed.org.

Obama, Barack. 2009. Inaugural—Address. Accessed June 15, 2015. https://www.whitehouse.gov/blog/inaugural-address/.

Piaget, Jean. 1985. *The equilibrium of cognitive structures: The central problem of intellectual development*. Chicago: University of Chicago Press.

Pittman, Karen. 2002. Balancing the equation: Communities supporting youth, youth supporting communities. *Community Youth Development Anthropology* 1(1): 19–24.

Prensky, Marc. 2010. *Teaching digital natives: Partnering for real learning*. Thousand Oaks: Corwin.

Selwyn, Neil. 2007. *Citizenship, technology and learning: A review of recent literature*. Bristol: Futurelab. Accessed June 15, 2015. http://www.futurelab.org.uk/resources/documents/lit_reviews/Citizenship_Review_update.pdf.

Skinner, Rebecca, and Christopher Chapman. 1999. *Service-learning and community service in K–12 public schools*. National Center for Education Statistics. http://nces.ed.gov/surveys/frss/publications/1999043/index.asp.

Stanton, Timothy, Dwight Giles, and Nadinne Cruz. 1999. *Service-learning: A movement's pioneers reflect on its origins, practice, and future*. San Francisco: Jossey Bass Publishers.

Sweency, Aldrin F., and Jeffery A. Paradis. 2004. Developing a laboratory model for the professional preparation of future science teachers: A situated cognition perspective. *Research in Science Education* 34(2): 195–219.

Tucker, M., Anne McCarthy, John Hoxmeier, and Margarite Lenk. 1998. Service-learning increases communication skills across the business curriculum. *Business Communication Quarterly* 61(2): 89–100.

Ward, Doug. 2015. Why change our teaching?" Faculty focus. Reprint from *Teaching Matters*, July 15, 2015. Accessed June 15, 2015. http://www.cte-blog.dept.ku.edu/why-change-our-approach-to-teaching/.

Warschauer, Mark. 2003. *Technology and social inclusion: Rethinking the digital divide*. Cambridge, MA: MIT Press.

Wenger, Etienne, Richard McDermott, and William M. Snyder. 2002. *A guide to managing knowledge: Cultivating communities of practice*. Cambridge, MA: Harvard University Press.

CHAPTER 6

Promoting an Interdisciplinary Campus Culture

Costanza Eggers-Piérola, Bonne August, Cinda P. Scott,
Pamela Brown, and Reneta D. Lansiquot

Abstract This chapter synthesizes best practices and lessons learned in order to facilitate the transfer of knowledge between courses, disciplines, programs, research fields, and industry. It describes strategic institutional integration that transforms approaches to recruitment, teaching, mentoring, supervision, communication, and collaboration within and across laboratories. Research and education are integrated with a focus on

C. Eggers-Piérola
Eggers International, Cambridge, MA, USA

B. August • P. Brown
Office of the Provost, New York City College of Technology, City University of New York, Brooklyn, NY, USA

C.P. Scott
Tropical Island Biodiversity Center, The School for Field Studies, Bocas del Toro, Panama

R.D. Lansiquot (✉)
English, New York City College of Technology, City University of New York, Brooklyn, NY, USA

© The Author(s) 2016   107
R.D. Lansiquot (ed.), *Technology, Theory, and Practice in Interdisciplinary STEM Programs*,
DOI 10.1057/978-1 137-56739-0_6

inquiry-based learning and developing a global workforce by expanding industry partnerships. This chapter also contributes to the dialogue on best institutional approaches focused on attracting, retaining, and preparing underrepresented students. The cross-institutional strategies, faculty development, and initiatives described provide real-life examples of what works toward these goals and what sustains and multiplies these efforts.

Keywords Inquiry-based learning • Case studies • Faculty development • Lab practice • STEM programs • Underrepresented students

Responding to a nationwide need to broaden the participation of underrepresented students in science, technology, engineering, and mathematics (STEM), in 2008, the National Science Foundation (NSF) initiated the Innovation through Institutional Integration awards. This effort sought to investigate how institutions could creatively integrate institution-wide efforts across NSF-supported programs and how such innovations could be sustained. In 2009, New York City College of Technology (City Tech), the designated college of technology of the City University of New York (CUNY), was a recipient of one of these seminal five-year NSF grants. *The City Tech I³ (Innovation through Institutional Integration) Incubator: Interdisciplinary Partnerships for Laboratory Integration* project designed and piloted professional development programs and college-wide initiatives to enhance STEM courses. The City Tech I³ strategy focused on increasing collaboration across departments, encouraging experimentation and curriculum change, and creating partnerships with community and industry mentors.

In essence, the City Tech I³ Incubator sought to provide innovative, real-life, hands-on experiences responding to the needs and learning styles of the diverse students that make up City Tech's student body. In the fall of 2014, City Tech's full-time undergraduate enrollment was 38.6 % Hispanic/Latino, 27 % Black, and 24.2 % Asian/Pacific Islander, with 57 % of its 17,374 students enrolled in associate programs, and 37 % enrolled in bachelor programs, most of these in science and engineering applied fields.[1]

PROGRAM DESIGN: THEORIES OF LEARNING AND THEORIES OF CHANGE

Since 1989, we have had a clear understanding of what contributes to effective higher education STEM programs:

- Learning is embedded in the community, is experiential and hands-on, and is meaningful to students, community, and faculty members
- Learning is a collaborative partnership between faculty and students
- Learning is rich in research[2]

In order to carry out these principles effectively for our students at City Tech, we needed both to increase communication between traditionally separate structures and practices and to create new collaborations both within the institution and between City Tech and external entities. Therefore, we engaged faculty in designing pedagogical change, especially through employing case study methods with interdisciplinary content[3]; we also forged external partnerships[4] and created apprenticeships and internships with community and industry partners in order to provide meaningful experiences related to concepts learned in STEM labs. Case study methods supply hands-on learning opportunities that allow students to explore real-life applications pertinent to their fields and industry needs.[5]

Key questions of the I^3 Project included the following:

- What program features contributed to the faculty's understanding and teaching of STEM?
- To what extent was a community of professional practice created to support educators' development and inquiry into best STEM teaching and learning practices?
- What infrastructures and partnerships were created that supported the effectiveness of the program?

Table 1 provides a model that identifies and defines the major needs, goals, and actions of City Tech's I^3 Incubator Project:

To generate systemic change, we used an approach that aligned common goals and metrics across all NSF supported projects. The City Tech I^3 project more than met its goals, transforming approaches to recruitment, teaching, mentoring, and supervision, as well as communication and collaboration within and across laboratories. To evaluate the program, we gathered data from multiple sources in order to address internal validity and triangulate findings. Participants who were surveyed across the five years included faculty members, department heads, and students from across STEM departments at City Tech. Most of the activities were documented separately in supplemental material available online.

Table 1   Theory of change model

Lab improvement	Means
Transform pedagogy: Move away from teacher-directed verification labs to incorporate authentic, inquiry-based methods and interdisciplinary content	Design new experiences for faculty Target junior and senior faculty Follow-up experiences with actions through committees and workshops Transform curriculum and lab manuals Improve coordination with adjuncts and coordinators
Enhance coordination:	Learning community is cultivated to ensure continuity and synchronicity between labs and lectures and provide opportunities for faculty interactions
Move away from outdated manuals and inefficient lab procedures to create current and engaging materials and labs for students	Cohesive College Laboratory Technician procedures and policies are planned Departmental surveys and input from faculty inform needed curricular changes
Motivate students: Move from technical, context-free, abstract content to personalized, real life application and compelling interaction opportunities for students	Involve students in authentic research Cultivate industry partners Involve students in discussing pedagogy Provide internship opportunities that engage students in meaningful community-based work
Develop state of the art learning experiences for faculty: Move from procedural and technical departmental meetings to cross-department sharing and action-planning	Bring significant training and support in case study and interdisciplinary methods Fund faculty results-based planning Foster external faculty development opportunities Create follow-up plans Involve and support committees and work groups
Motivate faculty:	Constantly disseminate and highlight faculty innovation
Move from segmented and top-down structures to ignite a sense of ownership and excitement in initiating new projects and methods	Increase time to share across departments Present and publish case studies and new interdisciplinary curricula broadly

The findings from these surveys and analyses of related documentation were presented in periodic formative evaluation reports and presentations in order to provide all stakeholders an opportunity to analyze the data as well as to contribute to the modifications called for in the I³ project's goals, methods, and plans. In this way, faculty and department heads used data to accelerate change and intentionally design sustainability.

IMPLEMENTING CASE STUDIES, INQUIRY-BASED LEARNING, AND PARTNERSHIPS

Project implementation took a two-pronged approach: providing support for faculty and developing a cross-institutional approach to enhance teaching and learning at City Tech. The first phases prioritized professional development, and the latter phases focused on institutional strategies that cemented and spread the knowledge, experiences, and skills relevant to hands-on teaching and learning, as is detailed in Table 2.

Case Study Model

The first effort in developing case studies concentrated on achieving a clear common understanding of case study methods and their use in interdisciplinary STEM labs. Three professional development workshops provided faculty and leaders from several departments' hands-on experience creating interdisciplinary case studies, and subsequently the case study model was implemented successfully in nine lab and lecture courses across nine programs. After the courses were offered for the first time, they were revised, and lessons learned were used to create more such courses.

Inquiry-Based Faculty and Student Experience

The City Tech I³ sponsored the Advancing Chemistry by Enhancing Learning in the Laboratory (ACELL) workshop, which helped foster the idea of complementing the case study approach with an intensive faculty/student inquiry experience. As a result, twelve faculty and twelve students worked closely together for two intensive days to discuss inquiry and cur-

Table 2 Ongoing data collection and data-driven planning of the implementation phases

Year 1	Year 2	Year 3	Year 4	Year 5
Development of surveys and needs assessment methods	Case study planning Student and faculty surveys	Case study professional development and planning Partnership-building Apprenticeships	Case study implementation Inter-disciplinary curriculum development Apprenticeships	Institutional strategies for inter-disciplinary studies, lab coordination and career prep

riculum change. The group identified the top characteristics of good labs and proposed changes that could be made to City Tech labs to incorporate key characteristics of inquiry and best lab practices. Subsequently, Biology and Chemistry faculty met to plan changes in lab manuals and practices in their departments. Additional support meetings and workshops covering related topics (e.g., grant writing and research, articulating goals and strategies, identifying common metrics, designing pre-lab assignments, and so on) bolstered these professional development efforts.

Partnerships

Cross-institutional partnerships that engage underrepresented students in work-related STEM activities are not only beneficial to students but also enhance the collaborating partners' diversity.[6] Hands-on, community-based internships that anchored student learning in real-life situations became an I^3 institutional strategy early on. The I^3-developed Anchors Internship Program, a pilot with the Brooklyn Navy Yard Development, changed the college's approach to career-ready, relevant learning experiences and led to subsequent internship programs benefitting more than 200 students each year.

INNOVATIONS AND INTEGRATIONS

The I^3 Incubator at City Tech proved to be an energizing and unifying force that helped shape the direction and mission of the College and has provided enduring results. The goals and objectives of the I^3 Incubator were distilled to the overarching themes, *Innovations, Integration,* and *Connections.*

Innovations refers to the advances and explorations in pedagogy and curriculum that reflect more relevant and hands-on experiences, as well as to processes that were consequential to refining and sustaining innovations.

Integration refers to achieving increased alignment and synergy among City Tech's NSF-supported projects through strategies that stimulate collaboration and sharing, as well as the design of systems, models, and templates to represent and facilitate integration.

Connections refers to developing meaningful and enduring partnerships within and outside the college to enable students to get state of the art, real-life application of the content.

Innovations in Ownership and Agency

"We really, really want to make changes," emphasized a faculty member who participated in I³ activities. Such change has to happen at each level, and therefore the I³ activities were directed at department chairs as well as junior faculty. Many times, junior faculty are more willing to make changes, but do not have the authority to do so and consequently may place themselves in a vulnerable position. The I³ activities have contributed to lessening these barriers by gathering a critical mass of professors across the College who feel supported and are working toward the same goals. As one faculty member comments, "You have to get to both of them"—meaning both administrators and faculty members—and went on to stress the positive influence of the energy of younger faculty and the need for support from administrators. The I³ project greatly benefited from the formation and inclusion of a program manager position which provided a link between faculty and administration.

Moreover, faculty need to feel a sense of ownership over innovations in programs and pedagogy, and this is created by informing faculty that there are goals that they can accomplish and meaningful changes they can make. Faculty need to be "involved in projects that help create the institution you want to teach in," I³ Principal Investigator, City Tech's Provost Bonne August, remarked. Even more vital is a passion for change. Faculty must desire a change, believe in it, and seek to bring it to reality through their teaching. Department heads, too, have to value change. The early I³ surveys brought compelling data to the department heads that spurred them to rally their departments. One department head remarks that these surveys gave faculty members "the taste of doing things" that propelled faculty members forward to shape a model of shared and distributed leadership across City Tech, a model which has been to be proven successful in increasing the capacity for change and improvement.[7]

Innovations in Collaborations and Interactions

Changing the nature of interactions in the lab was a pedagogical objective, but a corollary result was the enriched collaboration among faculty and between faculty and students. As a result of the I³'s targeted, hands-on professional development, faculty were inspired to make changes that offer students more decision-making and interaction opportunities. Even small changes play a role in transforming their teaching. One instructor describes the change in this way:

"Look at the instructions," I say, "but you guys decide what you are going to do, who does what." It's there, but I'm not telling them. The group activity, it's more planning. If you have an experiment, you have to plan. I didn't used to do this.

Another instructor remarks, "If we involve students a little more, we could learn a lot from them."

Qualitative data from Year 1 and Year 2 demonstrated that both students and faculty considered interactions and collaboration with other students and faculty beneficial in their labs. The faculty surveyed stated the "best thing" about the workshops were "Faculty-student teamwork. Everyone seems to be engaged. Students had the opportunity to give/propose ideas." According to the professors, students developed better relationships and communication with them, initiated activities, and were clearly motivated to do more case studies and real-life applications in their labs. Several instructors reported students would still come up to them, a year after taking their classes, and ask for more case studies.

For their part, students said that they "felt heard and enjoyed the exchange with faculty," and that the "best thing" about the workshops and cases was "working in groups and interacting with others." Further, the student said, "Hearing answers from different people will enable you to understand other people's perspective." They claimed they learned from each other, clarified doubts, compared outcomes, and had the opportunity to apply their expertise and knowledge.

Faculty appreciated the chance to get together and learn from each other. Some of the faculty were already collaborating on projects, but the I^3 sharpened the focus and promoted a common language and direction. New partnerships and collaborations sparked conversations that resulted in active planning and re-envisioning existing infrastructures and initiatives, such as the Year 5 workshop given by the Interdisciplinary Committee to support proposals for team-taught interdisciplinary courses.

Innovations in Pedagogy

Faculty who participated in I^3 described themselves as feeling more motivated to innovate as a result of the pedagogical discussions arising from the workshops and planning meetings. As one faculty member put it, "once exposed to inquiry-based, authentic teaching and learning, and inquiry approaches, many are excited and willing to incorporate them," adding that the most important criterion to judge an innovation is whether it "motivated and educated students to further actively pursue STEM investigation." Indeed, the level of enthusiasm and camaraderie was palpable. One faculty member commented, "Using the case study in the classroom for me is a new way to teach, quite exciting," and another said in a focus group, "Before we worked together, we hated each other" sparking laughter from the rest of the group. Another faculty member elaborated the benefits in this way:

> In talking with students who participated, they really liked getting the perspective of the professors and felt more connected to the education.... [They said,] These professors care about what we are doing. I have seen them do this, now I understand what they are trying to do in the classroom.

Innovations in Curriculum Changes

Two major changes in the way curriculum is shaped across City Tech were activated by I^3:

1. Optimization of lab manuals and coordination.
2. Integration of interdisciplinary and case study content into STEM programs.

Lab manuals dictate what happens in the labs and in final exams. Biology lab manuals were updated, revised, and made more visually useful with color illustrations. Questions were added to involve students in critical thinking. In addition, instructors began piloting new hands-on kits obtained through their own grant writing efforts in order to complement the mitosis/mycosis labs described in the manual. Others have started to employ online platforms and databases to give students more hands-on, real-life experiences and chances to engage in research.

Moreover, course curricula in departments across the college were revamped and made current through a multimodal approach, customized by each faculty member to suit the possibilities and needs of their current and proposed courses. Case studies were designed by fourteen faculty members as a direct result of the I^3 and incorporated across a variety of departments. Interdisciplinary case studies fulfill a new General Education requirements detailed below under the heading, *Integration through changes in infrastructure.* These new interdisciplinary courses integrate material and approaches from two or more disciplines as well as writing skills, making the cases interesting and memorable to students. Original and newly developed case studies and interdisciplinary courses include such interesting and dynamic collaborations as the case study "Mr. Rump Goes to Washington" (Electrical Engineering Technology and Environmental Control Technology) and courses "Learning Places: Understanding the City" (Architecture Technology and Library Science) and "Energy Resources" (Chemistry and Construction Management and Civil Engineering Technology).

Further highlights of the impact of I^3 on curriculum and pedagogy include the following:

- In the Mathematics department, an instructor was able to incorporate case studies in an extracurricular, informal format.
- A faculty member in the Chemistry department has designed a wiki to increase active student participation, specifically inviting them to input and debate their own definitions of STEM terms.
- Pre-lab workshops with real-life examples were modeled and encouraged.

Integration Through Institutional Changes

City Tech was founded after World War II as an associate degree-granting community college offering workforce-oriented programs, but since the 1980s has been steadily increasing its baccalaureate programs. Now City Tech offers 24 baccalaureate and 26 associate programs in applied and career-focused studies and is preparing to accommodate its rapidly increasing student population by building new facilities and taking a cohesive approach to improving pedagogy and internships.

The most significant efforts in establishing the foundation for institutional integration and transformation began with a needs assessment that found that too often STEM students at City Tech were offered rote rather than inquiry experiences. In response, one effort focused on reading and writing assignments in Biology and Chemistry courses, which were revised to engage students with narratives and interesting Pre-Lab assignments that prepared students to employ applied knowledge and critical thinking skills. Instructors also changed their grading formulas to place value on these critical skills, a small but key strategy.

Integration Through Changes in Infrastructure

The team-teaching aspect of the City Tech I³ initiative, which has proven to be vital, presents the challenge of adjusting scheduling and space in a very overcrowded institution. This not only involves finding periods when teachers can plan and teach interdisciplinary case studies and courses together, but also adapting the design of a new building to accommodate the new lab instruction and technology requirements.

One advantage to so much collaboration is seen in the science departments, which must coordinate supplies and lab equipment. The I³ faculty made strides in this direction by suggesting a clear set of protocols, which included trimming some bureaucracy and redefining the job description for the College Laboratory Technician to stress the need to coordinate and find different ways of working and implementing systems that will support pedagogical changes.

City Tech's Interdisciplinary Committee has been instrumental in institutionalizing the interdisciplinary and team-teaching approaches explored through the I³. Among other things, this committee created an institutional definition and criteria for an "interdisciplinary course" and identified clear learning outcomes for students. This team worked effectively and achieved its goals because members worked on other committees together. As the committee founder, Dr. Reneta D. Lansiquot, explains, "We, the Interdisciplinary Committee, we're part of the Gen Ed Committee...and thus we have 'infiltration.'" As a result, the committee knew how to collaborate, create programs, and gain support for them.

While the Interdisciplinary Committee began this work, the General Education Committee's decision to make a liberal arts and sciences interdisciplinary course a part of the new General Education Common Core requirements was critical to institutionalizing interdisciplinary studies.

Integration Through Synergistic Activities

It is clear that these efforts are not simply fortuitously synergistic, but rather are now being consciously and strategically aligned thanks in great part to the I³ grant. As one faculty member comments,

> With new leadership in the department, and the kind of support we've been able to give the chair and the junior faculty, all of this is really essential to moving this development forward. Now they're talking about developing

new programs, new labs, much more streamlined systems, and really this came about because I³ was the catalyst.

Successful subsequent NSF-funded grants at City Tech, including Research Experiences for Undergraduates (REU), Opportunities for Enhancing Diversity in the Geosciences (OEDG) grants, and the Advanced Technological Education (ATE) Fuse Lab grants, built on I³ principles and practices, such as its focus on authentic research with real-life application. Institutional investment in student research has increased dramatically: The Undergraduate Research Program, which in 2006 had twenty students, now has five hundred. A current example is a grant from Cold Spring Laboratory for professional development on methods of introducing DNA sequencing into the curriculum. Another authentic research experience led by an I³-participating instructor involves collecting samples of bacteria from subway handrails and other public places, combining all the data, and mapping out the different bacterial profiles where students live. Changes in lab technology were enabled through complementary College funding. One faculty member observes,

> Previously, students were only able to carry out verification labs, and now students are doing discovery by going online, seeing 3-D representations of molecules, simulations of reactions, all things they hadn't been able to do before, and all this was prompted by faculty receiving those initial surveys.

MAKING CONNECTIONS

Industry Relevance

The Year 2 faculty lab surveys identified industry relevance and mutually beneficial relationship with industry as the major challenges in transforming labs. Since then, a more guided and intentional process for paid internships has been cultivated through I³. Before the grant, most internship opportunities came about by departments or individual professors who had relationships in industry and labs. Prompted by the success of the Anchors Internship Program begun through I³ with the Brooklyn Navy Yard Development Corporation and continuing through the subsequent cohorts of internships developed through the Brooklyn Tech Triangle Internship Program, and the CUNY Service Corps, the nature of apprenticeships and partnerships was transformed. The CUNY Service Corps has already enabled nearly 400 students to obtain paid year-long service experiences in non-profit and governmental agencies. In addition, the Brooklyn Tech Triangle Internship Program has placed well over 100 students in paid internships in technology firms in the burgeoning technology hub in the area surrounding City Tech.

These partnership efforts are reaping mutual benefits. Recently, Dr. Russell Hotzler—the President of the College—attended an event at the New York City Office of Small Business Services. Dr. Hotzler described it as a "love fest" of appreciation for the students and the College:

> The message from the employers was, again and again: "I never heard of City Tech before, but these students...have completely blown my mind. I want more of them!," because the students not only have the skills from their discipline but they were prepared for the internship.

Student preparedness for internships and work with industry have benefitted from the I³'s pedagogical approach, about which, one faculty members comments, involves "learning how science is done" and "learning how scientists think" and more generally "putting in place habits of mind that people will continue to develop throughout their careers."

Internal Connections

What aids establishing meaningful internal connections? According to participating faculty, aside from opportunities to develop and meet together, three conditions are paramount: Size of department, position and power to enact changes, and consistency. Another lesson learned is that having consistent messages and representation of I³ members across college-wide committees aids in shaping and acting on innovation plans, as faculty who have been actively involved in I³ are working from a common base of understanding.

Building a Common Culture

In essence, the I³ Incubator represents a coalescence of previous efforts to improve STEM teaching and learning across the College and to foster the adoption of a teaching philosophy that will inspire and support the persistence of a diverse and increasing body of students in

STEM studies and careers. One of the most overlooked and vital influences on integration efforts is the development of a common language. "Authentic research" is still the most challenging word to agree on across departments. Dr. Pamela Brown describes her interpretation of the term:

> The signature characteristics of authentic research experiences are that the results are not known before the experiment begins and that there is a potential for creating or collecting new data or knowledge. Whether it worked or not, it is the potential.

GOING FORWARD

Leadership is essential to sustaining a system-wide initiative. To be effective, someone in a leadership position cannot dictate change to others, but must create conditions and make tools accessible that inspire and support others. Thus, supporting, not imposing, is essential, otherwise buy-in will be thwarted and results will be limited.[8] The I³ faculty's investment and energy are evident as they continue to generate solutions and implement pedagogical changes on their own initiative. Strategic proposals bring together and transcend departments, such as the recently funded National Institute of Health Bridges to the Baccalaureate grant to partner with Brooklyn College. City Tech students in Psychology, Chemistry, and Biology associate programs will receive research opportunities and other support that will engage participants, promote academic success, and enable them to transfer successfully to Brooklyn College with a good background in research and solid preparation.

Going forward, the strong internships and partnerships with industry will scaled up, these thanks in part to convergent needs in the community. Provost August observes,

> We've been really lucky in that around us all these opportunities have sprung up lately...there's the Brooklyn Tech Triangle, the Dumbo complex...so our location, which was the road to nowhere for a really long time, turned out to be very important to IT and advanced manufacturing and other kinds of things happening in downtown Brooklyn. It's not only that it is here, but that we are ready for it from the kinds of work we've done with I³ developing the initial internship with Brooklyn Navy Yard Corporation, to understanding everything that needed to be in place for our students to be ready to take these experiences.

The interdisciplinary approach will be maintained by two important structural changes: (1) The General Education Common Core requirement of an interdisciplinary course for baccalaureate students and (2) the Interdisciplinary Committee's support for faculty development of special topics courses taught by interdisciplinary teams. Each semester, faculty will propose, develop, and plan new team-taught courses.

In addition, the case study methodology will soon be further developed to improve students' skills in using sequencing, selection, and repetition structures in computer programming. Dr. Reneta D. Lansiquot and Dr. Candido Cabo are leading an effort in which, faculty who teach Problem Solving with Computer Programming, and five students who successfully completed their "Story-Telling in Action-Adventure and Role-Playing Games" First-Year Learning Community course were involved. Students wrote case studies with Drs. Lansiquot and Cabo, then served as a focus group during the case study development. Last year, the faculty implemented the case studies to incorporate narrative elements in all sections of the course. As one faculty member comments, "The case study is a wonderful pedagogical tool created as an interdisciplinary effort," adding that the case study is "a great assignment for students and it's a great way to look at issues that transcend the disciplines and bring multiple perspectives on an issue or on a problem."

CONCLUSION

In the end, the five years of the I^3 Incubator not only achieved its end goals, but also has yielded unanticipated action results that have proliferated throughout the College. I^3 activities have clearly brought together a series of elements that stimulated change, capacity, and collaboration among the faculty and between faculty and students, as well as set the stage for departmental and institutional restructuring to enhance teaching and learning experiences across STEM and other departments. In effect, these aspects encompass the main aim of an institution-wide incubator: To sustain and proliferate a community of inquiry and a space for innovation and integration within the institution.

We feel it is important to end the story of the I^3 grant at City Tech by discussing the key aim of the Broadening Participation grants of which the I^3 has been a part: Addressing the inequities in accessing STEM careers. City Tech, as a Black- and Hispanic-serving institution, strives to extend its founding mission to provide a wide variety of students not only with associate degrees but also incentives to continue their studies and careers. City Tech has now proven itself as a baccalaureate degree-granting institution that attracts increasing numbers of diverse students to advanced studies and careers in STEM. As of 2014, the number of students receiving baccalaureate degrees exceeds the number receiving associate degrees.

Furthermore, the college has transformed many of its systems and ways of working to ensure that this pathway is accessible to more and more students. The initiative has launched a culture of community, making STEM education more interactive, more social, more relevant, and more empowering for both students and faculty. The building of this community is what makes women and underrepresented diverse students feel comfortable and interested in STEM, as well as taking part in meaningful real-life experiences that solve problems in their communities.[9]

The City Tech I[3] Incubator project has positioned the College to respond to its growing population of diverse students by remodeling its programs, innovations, and infrastructures using an integrated, cross-departmental approach and a College-wide attention to partnerships. Relationships were forged with a large number of external partners to cement the connections the students made to meaningful work and contacts in career fields, and enabled students to develop a professional identity, all key to attracting and keeping underrepresented students in STEM fields of study. More importantly, perhaps, these innovations have helped us prepare our students more fully for the challenges they will face in the twenty-first century.

NOTES

1. For the most current enrollment information, see http://air.city-tech.cuny.edu/air/FRD.aspx
2. Project Kaleidoscope, *What works: Building Natural Science Communities*, ed. Jeanne L. Narum (Washington, D.C.: Author, 1991); Sigma Xi. *An Exploration of the Nature and Quality of Undergraduate Education in Science, Mathematics and Engineering: A Report of the National Advisory Group of Sigma Xi, the Scientific Research Society* (Racine, WI: The Scientific Research Society, 1989).
3. Marilyn J. Amey, Denise Brown, and Lorilee R. Sandmann, "A Multidisciplinary Collaborative Approach to a University-Community Partnership: Lessons Learned," *Journal of Higher Education Outreach and Engagement 7*, no. 3 (2002): 19–26; Cinda

P. Scott, Bonne August, and Costanza Eggers-Piérola, "All Hands on Deck: Using Case Studies to Support Institutional Change." In *Cases on Interdisciplinary Research Trends in Science, Technology, Engineering, and Mathematics: Studies on Urban Classrooms*, ed. Reneta D. Lansiquot, 320–348 (New York: Information Science Reference, 2013).

4. Hanover Research, *Project-Based Learning and Best Practices for Delivering High School STEM Education* (Arlington, VA: Hanover Research, 2015).

5. Aman Yadav, Megan Vinh, Gregory M. Shaver, Peter Meckl, and Stephanie Firebaugh, "Case-Based Instruction: Improving Students' Conceptual Understanding Through Cases in a Mechanical Engineering Course," *Journal of Research in Science Teaching* 51 (2014).

6. Beatriz Clewell, Clemencia Cosentino de Cohen, Lisa Tsui, and Nicole Deterding, *Revitalizing the Nation's Talent Pool in STEM: Science, Technology, Engineering and Math* (Washington, D.C.: The Urban Institute, 2006).

7. Alma Harris, *Distributed Leadership Matters* (Thousand Oaks, CA: Corwin, 2013).

8. Allan Bolton, "A Rose by Any Other Name," *Quality Assurance in Education* 3, no. 2 (1995): 13–8.

9. Harvard Graduate School of Education Askwith Forum, "A Space of Their Own? Girls, Women, and STEM" (Cambridge, MA: Harvard Graduate School of Education, 2015).

Acknowledgments The authors thank the faculty, students, and staff of City Tech who have generously contributed to our understanding of the project's impact. We also wish to thank Dr. Russell Hotzler, President of City Tech, and our Program Directors at NSF for their support. This material is based upon work supported by the National Science Foundation under Grant No. 930242. Any opinions, findings, conclusions, or recommendations expressed in this material are those of the authors and do not necessarily reflect the views of the National Science Foundation.

BIBLIOGRAPHY

Amey, Marilyn J., Denise Brown, and Lorilee R. Sandmann. 2002. A multidisciplinary collaborative approach to a university-community partnership: Lessons learned. *Journal of Higher Education Outreach and Engagement* 7(3): 19–26.

Bolton, Allan. 1995. A rose by any other name. *Quality Assurance in Education* 3(2): 13–18.

Clewell, Beatriz, Clemencia Cosentino de Cohen, Lisa Tsui, and Nicole Deterding. 2006. *Revitalizing the nation's talent pool in STEM: Science, technology, engineering and math*. Washington, DC: The Urban Institute.

Hanover Research. 2015. *Project-based learning and best practices for delivering high school STEM education*. Arlington: Hanover Research.

Harris, Alma. 2013. *Distributed leadership matters*. Thousand Oaks: Corwin.

Harvard Graduate School of Education Askwith Forum. 2015. A space of their own? Girls, women, and STEM." Cambridge, MA: Harvard Graduate School of Education. http://www.gse.harvard.edu/news/15/02/space-their-own-girls-women-and-stem.

New York City College of Technology. 2014. *Frequently requested student data*. Brooklyn: New York City College of Technology. http://air.citytech.cuny.edu/air/FRD.aspx.

Project Kaleidoscope. 1991. In *What works: Building natural science communities*, ed. Jeanne L. Narum. Washington, DC: Author. http://files.eric.ed.gov/fulltext/ED351188.pdf.

Scott, Cinda P., Bonne August, and Costanza Eggers-Piérola. 2013. All hands on deck: Using case studies to support institutional change. In *Cases on interdisciplinary research trends in science, technology, engineering, and mathematics: Studies on urban classrooms*, ed. Reneta D. Lansiquot, 320–348. New York: Information Science Reference. doi:10.4018/978-1-4666-2214-2.ch013.

Sigma Xi. 1989. *An exploration of the nature and quality of undergraduate education in science, mathematics and engineering: A report of the national advisory group of Sigma Xi, The Scientific Research Society*. Racine: The Scientific Research Society.

Yadav, Aman, Megan Vinh, Gregory M. Shaver, Peter Meckl, and Stephanie Firebaugh. 2014. Case-based instruction: Improving students' conceptual understanding through cases in a mechanical engineering course. *Journal of Research in Science Teaching* 51: 659–677. doi:10.1002/tea.21149.

INDEX

© The Author(s) 2016
R.D. Lansiquot (ed.), *Technology, Theory, and Practice in Interdisciplinary STEM Programs*,
DOI 10.1057/978-1-137-56739-0